THE MARCONI BEAM WIRELESS STATIONS OF SOMERSET

by
LARRY BENNETT

ISBN: 978-1-80369-785-7

www.newgeneration-publishing.com

 New Generation Publishing

CONTENTS

Front Cover Photograph: A general view of the Bridgwater Beam
Wireless Station showing the aerials for the Canadian and South African
circuits.

Rear Cover Photographs:
1. Somerton Wireless Station, external view.
2. Somerton Wireless Station, bank of receivers.

INTRODUCTION

The County of Somerset has been blessed with numerous pioneering radio stations over the years; the world's largest maritime radio station (Portishead Radio) had its control centre at Highbridge, with transmitters located further up the Bristol Channel at Portishead. The first radio transmissions over water took place from Brean Down; an experimental station was established at Chedzoy; a wartime direction-finding station was set up at Stockland Bristol, near Bridgwater; and the Marconi 'Beam' Wireless Service established receiving stations at Bridgwater and Somerton. In addition, the BBC operated many broadcasting transmitter sites in the county, notably at Washford, Clevedon and Pen Hill, Wells.

The 'Beam' service was an early service designed to link countries (and indeed, continents) with each other by the use of highly-directional radio aerials, using high-speed telegraphy, telephony and facsimile circuits.

One such station was located at Huntworth, just outside of Bridgwater, which was the receiving station for the Canadian and South African services, with the associated transmitting station being located at Bodmin in Cornwall.

The Bridgwater station was only active for around 13 years, being replaced by a larger receiving station at Somerton, but it was a vital part of the development of communications history, and one which deserves to be recorded. The aerials have long since been removed, but some of the original buildings still exist, being converted into both private accommodation and commercial use.

Those who pass the M5 junction South of Bridgwater will today struggle to imagine the vast aerial masts and wires which once dominated the skyline, and which provided a valuable link to two continents.

In addition, recently-found information about the experimental site at Chedzoy, on the other side of Bridgwater, makes for fascinating reading. This station was only active for a couple of years but its role in the history of long-distance radio communication cannot be underestimated.

The Somerton receiving station, which took over from Bridgwater in the late 1930s, is also covered in depth. This station became the receiving site for the high-frequency maritime radio service in the early 1980s.

Brief overviews of the Bodmin and Dorchester transmitting stations (sister stations to Bridgwater and Somerton) are also included for the sake of completeness.

It is also useful and relevant to include the origins and political implications of the early days of the complete service, which serve as a background to the development and operation of the stations at Bridgwater/Bodmin and Somerton/Dorchester. This is featured in Chapter 1, and is followed by an overview of the UK network which handled the global circuits for many years. Use of Parliamentary archives and Hansard extracts have been used to emphasise the political and financial implications of the service, which give a fascinating insight into the background over which the stations were established and operated.

Many photographs and newspaper articles have been sourced, with some information being published for the first time. Some photographs have been scanned from the original newspaper articles as the original photographs are no longer available.

The story of these stations in Somerset is full of interesting facts, both technical and anecdotal, and in these days of instant global communication it is indeed interesting to look back at the early days of wireless communication and the sense of wonderment it evoked.

Larry Bennett
February 2023

ACKNOWLEDGEMENTS

- The family of the late Phil Lewis, whose collection of information about the Bridgwater Beam Wireless Station has formed the basis of this book;

- Paul Hawkins, for his welcome assistance and whose book 'Point to Point' was an invaluable reference throughout my research, and is recommended further reading;

- The editors of 'Radio User', 'Practical Wireless' and 'Wireless World' who kindly gave me permission to use extracts from their archive magazines;

- Alan Hartley-Smith, who gave kind permission to use extracts from the Marconi Archive;

- John Weeks of Chedzoy, who has assisted greatly with my research into the Chedzoy Wireless Stations;

- J. R. McCullum and J. D. Sharples, whose history of Bodmin Wireless Station has been used as a point of reference to the station;

- The late David Bentley, a former colleague of mine at Portishead Radio, for his photographs of Somerton Radio from 1999;

- The late Ramsay Stuart, another former colleague at Portishead Radio, for his memories and photographs of Somerton Radio;

- BT Heritage & Archives, whose assistance has proved essential to the project;

- The Bridgwater Heritage Group, for their valued assistance and support;

- Eugene Byrne, for his assistance with newspaper archive research.

- Hansard extracts are used in this publication which contains Parliamentary information licensed under the **Open Parliament Licence v3.0.**

FURTHER READING

- Portishead Radio – Larry Bennett, New Generation Publishing, 2020

- All Ships, All Ships – Larry Bennett, New Generation Publishing, 2021

- Point-to-Point – Paul Hawkins, New Generation Publishing, 2017

- Dorchester Radio – Paul Hawkins, privately published, 2004

- The Marconi Beam Wireless Stations of Essex - Paul Hawkins & Paul Reyland, New Generation Publishing, 2022

- Rugby Radio Station – Malcolm Hancock, privately published, 2016

CHAPTER 1

THE ORIGINS OF THE BEAM WIRELESS NETWORK

Before the days of the internet and instantaneous global communications, one of the ways in which messages could be sent or received quickly between continents was by the use of radio (or wireless, as it was usually known in the early days). Communication by cable was of course also available but in the early days this could be expensive and sometimes unreliable.

To complement the radio service, numerous stations were constructed with large aerial systems directed at the required country. These were known as 'beam' aerials, which 'beamed' their signals towards their target, ensuring greater transmission efficiency and reception quality.

The designer of these aerials was one Charles Samuel Franklin (1879-1964) who also devised the variable capacitor, ganged tuning, variable coupling, co-axial cable and the 'Franklin' Oscillator, and may be regarded as one of wireless's great pioneers. He worked for his entire life at the Marconi Company and was responsible for many of their developments in this field.

Charles Samuel Franklin (1879-1964).

In its simplest form, messages could be sent and received (in the UK) via the Post Office network and through the Central Telegraph Office (C.T.O.) in London. A network of transmitting and receiving stations was established throughout the UK, their locations being carefully selected for optimum efficiency and lack of local interference.

It was as long ago as 1902 that the seeds of international wireless telegraphy were sown, when it was reported on 12th April that:

"The Marconi Wireless Telegraph Company on Friday night issued the text of the agreement with the Canadian Government. The company agrees to erect stations in the

United Kingdom and Canada. The Government agrees to pay 80,000 dollars to the company for the erection of the station in Nova Scotia. The company agrees, if its operations prove successful, to transmit general messages at fully 60 percent less than the rates now charged for cablegrams. That is, the company will not charge more than ten cents a word. Government messages and press messages will be sent for not more than five cents a word."

Sr. Guglielmo Marconi (1874-1937).

Early radio experiments took place on long wavelengths where the power of the transmitter was thought to be the most important aspect. However, the tests conducted on short wavelengths proved that long-distance communication could reliably take place with much lower power, should the optimum time and frequency be selected. Confirmation of such tests were provided by radio amateurs, who proved that communication around the world could be effected using low power and basic aerial systems.

Initial plans for some sort of international wireless communications network were discussed prior to the outbreak of war in 1914, but these were shelved for the duration of the conflict.

At the cessation of hostilities in 1918, the British Government thought it desirable to link up the countries of the Empire using wireless telegraphy to complement the existing cable service, and the distances obtained by the radio amateurs on short-wave were of considerable interest. Of course, many official wireless stations had been operating for many years on long and medium wavelengths, but these tended to be of short or medium-range only, maybe up to 250 miles during the day and up to 2,000 miles in darkness.

With regard to the actual development of the Beam Stations themselves, it was the pioneering experiments undertaken by radio amateurs which demonstrated the properties of long-distance communication using short-wave frequencies. In December 1921, the first Trans-Atlantic communication on a wavelength of 200 meters took place, with signals being received in Scotland on a 'Beverage' aerial, especially constructed for that purpose.

In 1923, two-way communication at night was established between the USA and France and also between the USA and England on a 100 metre wavelength, and this was followed in May 1924 when communication between Argentina (using 120 metres) and New Zealand (using 90 metres) took place.

The honour of the first station in the 'Imperial' wireless chain was set up in 1921, using a site at Leafield,

Oxfordshire, although it did not use the 'beam' principle of radio transmission. The Post Office made the most of publicising the event, and details were promulgated via the national press on 19th August 1921:

"Mr. F. G. J. Kellaway, the Postmaster-General, at Leafield, Oxfordshire, yesterday attended the completion of the first station in the Imperial wireless chain, and incidentally pictured the time when the Prime Minister in London would be able to talk direct to Mr. Hughes in Australia.

The Leafield station, which has been designed entirely by the Post Office, will communicate with the corresponding station at Abu Zabal, near Cairo. This station is being proceeded with rapidly, and is expected to be ready in three months' time. It will also be used for communication with Mesopotamia, and probably for broadcasting news to India. The rate by wireless between this country and Egypt will be 9d, as compared with the present charge of 1s. Leafield and Abu Zabal will form the first pair of a series of four stations, the third being in East Africa, and the fourth in South Africa. In accordance with the proposals of the Imperial Wireless Telegraphy Committee, another pair of stations will be erected in England and Egypt, and these will be continued to India, Singapore, Australia, and Hong Kong.

Besides the Postmaster-General, there were present at the ceremony Sir Thomas Williams and Mr Blackmore, members of the Post Office Advisory Committee; Admiral of the Fleet Sir Henry Jackson (chairman) and Colonel Cusins (W. O. Signals), of the Radio Research Board; Major Purvey, Assistant Engineer-in-Chief of the Post Office;

Mr E. H. Shaughnessy, head of the Wireless Telegraph Department; and Mr F. J. Brow, Assistant Secretary in Charge of Telegraphs.

An interesting part of the ceremony was the sending of the first messages from the station. These were in the following terms:

> The Postmaster-General sends greetings to all British stations within range of the occasion of completion of this the first station of the Imperial wireless chain, and trusts that the station will help to knit still closer the bonds which bind together the different parts of the Empire.

> The Postmaster-General sends greetings to all European stations and to other foreign stations in range on the occasion of the completion of this the first high-power station owned by the British Post Office, and trusts that the development of wireless communication will help to knit still closer the bonds of unity which bind the British Empire to all other States.

Within half an hour replies were received from Malta, Paris, Christiania, Posen, Prague, Denmark, Budapest, Rome, and Berlin. The last-named, in sending congratulations, said "the signal was received here on a clear and good note."

The Under-Secretary of the French Posts and Telegraphs said the new station would "contribute greatly to the maintenance of our cordial relations, and also to the development of the world's wireless communications"

The Postmaster-General, in an address, said it was felt that such an historic occasion should be marked by some public ceremony. The Post Office suffered from too much modesty, and unless attention was occasionally called to some of its developments the public was apt to regard it as a slow and dead-alive body. Those who had visited that undertaking had had a real tonic experience. What particularly appealed to him was that all the principal work was of British production and the result of British inventions. *(Cheers)*.

It was almost impossible to believe that wireless telegraphy was only invented in 1886, when Mr. Marconi, then a young man of 22, visited England and took out the first wireless patent. In 1908 the Post Office, which had been made the statutory authority for wireless in 1904, first entered the field of commercial wireless. Now the Post Office coast station service was believed to be the most efficient in the world. At Northolt, a medium-power station was being erected which would communicate with Central and Eastern Europe. At the Imperial Conference great interest was taken by the Prime Ministers in the possibilities of wireless telephony. No one could dogmatise on the possibilities of wireless telegraphy and telephony. One of the drawbacks was its publicity, but he was confident that the brilliant and audacious men with this art in their hands would find some way of solving the difficulties. He believed the undertaking which they were inaugurating would do a great deal to bind closer together the different parts of the Empire. The wireless chain would link almost instantaneously almost every part of the Empire. It had been a subject of controversy as to whether it was better to have stations covering a moderate distance of about 2000 miles or to set up stations of greater power, which would give communication between England and Australia.

But the grandiose people with grandiose schemes always broke down the moment you came to the question of expense. In increasing power you increased the expense out of all proportion. By the 2000-mile stations they had one of the most reliable, economical, and, taking the year over, the speediest of systems.

"We are in the presence of one of the most baffling and fascinating of subjects," added the Postmaster-General, "but we have been able to harness this mysterious force, and use it for linking up the families of the Empire. It is an achievement of which we have a right to be proud, and of which the British Post Office has a right to be proud" *(Cheers)*.

The station officially opened on 24th April 1922.

A committee headed by Sir Henry Norman that same year was established to look into expansion of the network, which was to link up the countries of the Empire in stages of up to 2,000 miles, using the tried and tested high-powered long wave frequencies; stations in Egypt, Kenya, South Africa, India, Singapore, Hong Kong, Australia and Canada were discussed, which would have meant a message for Australia would have to have been relayed via stations in

Egypt, India and Singapore before arriving in the country of destination.

The Dominions, however, were not convinced and preferred the option of installing larger stations and communicating directly with England. A Parliamentary Commission agreed, and the G.P.O. commenced the construction of a suitable station at Rugby, which would operate with a power of 1,000 kW, using 12 masts of 800 feet high, and operating on a wavelength of around 18.000 metres. This would cost in the region of £350,000.

The Commission reported that:

"Mr. V. H. Eccles, the vice-chairman of the Wireless Telegraphy Commission, in an explanatory summary attached to the report of the Commission, stated that the Commission recommended:

Thermionic valve stations should be erected in England, Canada, South Africa, and India, and one station also in Egypt to duplicate the arc station now being completed by the British Post Office at Abu Zabal.

At each of the other centres, namely East Africa. Singapore, and Hong Kong, an arc station should be erected with space for the addition of thermionic valve plant later.

The average cost of the overseas stations will not exceed £100,000 exclusive of the cost of erecting residences for the staff at some of the stations. The stations in England, Egypt, East Africa, Singapore, and Hong Kong, for which the Imperial Government is presumably responsible, should not exceed in the aggregate the sum of £853,000.

It is emphasised that during portions of each day much of the Imperial strategic, official, and news traffic, carried on by direct communication between any pair of principal centres, the intermediate stations being omitted, but the intermediate stations would be necessary for relay work during the less clear portions of the day, and also for handling their own local traffic.

The news messages transmitted from the principal centres could be received at many stations in the Empire, for example in New Zealand, at the cost of an inexpensive addition to their existing receiving equipment. Foreign stations in many parts of the world would be able to pick up news and propaganda from one or other of the principal centres."

The Leafield station, 1922.

In the meantime, the authorities in Australia, South Africa, Canada and India negotiated contracts with the Marconi Company to install high-powered stations to complement and communicate with the new station at Rugby.

Such understanding was used to great effect by the Beam Wireless Service, details of which can best be explained by an excellent article written in 1926, which gives a clear overview about the origins and purpose of these stations:

"One of the most remarkable developments in wireless telegraphy within recent years has been the discovery of the merits of the short-wave system for long-distance communication. Very short waves, under a metre in length, were used by Hertz in his pioneer experiments which proved the existence of wireless waves, but in the practical application of wireless to commercial purposes, the possibilities of short waves were largely neglected, and as the range of communication increased the trend of development took the direction of using longer and longer waves and greater and greater power, reaching its climax in giant stations, such as Rugby, which employ wavelengths of the order of 20.000 metres and power of the order of 1,000 kilowatts.

During the First World War, Senatore Guglielmo Marconi, with the assistance of Mr. Franklin, took up the study of 'directive' telegraphy, using reflectors to concentrate and direct the radiated waves in a narrow beam, after the manner of a searchlight. To secure effective results, it is found that the dimensions of the reflector must be of the same order as the wavelength employed, and this consideration precludes the practical use of reflectors for all but very short wavelengths. The development of the short wave system was facilitated by the evolution of the valve transmitter which solved the problem of producing very short waves. Up to this stage it had been assumed that the range of short waves was too small for practical purposes, but in 1921 it was discovered that short wave stations, using very low power, were capable of communicating across the Atlantic.

Numerous tests between stations and Marconi's yacht *'Elettra'* took place throughout the early 1920s, using various frequencies at different times of day in order to comprehend radio propagation characteristics. Results of the tests were examined in great depth by mathematicians and radio engineers to ensure that radio paths could be used at the most efficient times and on reliable frequencies.

Marconi's yacht 'Elettra' circa 1923.

This discovery was followed up by Senatore Marconi, who established the practicability of world-wide communication during certain hours of the day on waves under 100 metres in length, and showed that by means of reflecting screens the radiation could be concentrated within a narrow beam, thus ensuring greater economy in power and freedom from interference.

Up to this date it had been the policy not only of the Imperial Government, but also of the Dominion and Indian Governments, to provide Imperial wireless communication by means of high power stations, but suddenly in 1923, Senatore Marconi disconcerted all preconceived notions on the subject by announcing the wonders of the short wave directive - or so-called 'beam' system - which he claimed would provide adequate services for a limited number of hours per day at a much smaller capital expenditure.

The Government decided to give the system a trial, and an agreement was accordingly made with the Marconi Company in July 1924, for the erection on sites to be provided by the Government of a beam station in this country for communication with a reciprocal station in Canada, which they undertook to erect through their affiliated Canadian Company, with provision for its extension, if required, for similar services with South Africa. Australia and India. The Governments of South Africa, Australia and India adopted the same policy and arranged for the provision of reciprocal beam stations.

The Marconi Company, however, subsequently came to the conclusion that, for technical reasons, the original scheme of concentrating all the sending stations on one site and all the receiving stations on another was impracticable, and a Supplementary Agreement was concluded in November 1924, which provided for the erection of two groups of two stations each, one in the south-west of England for communication with Canada and South Africa and the other in the Eastern Counties for communication with India and Australia.

Considerable difficulty was experienced in obtaining sites which would satisfy the technical requirements, but ultimately sites for the sending and receiving stations for communication with Canada and South Africa were secured at Bodmin and Bridgwater respectively, and sites for the sending and receiving stations for communication with India and Australia at Tetney (near Grimsby) and Skegness respectively. The corresponding stations in the Dominions are situated near Montreal, near Cape Town, near Poona and near Melbourne. The Agreement provides that the sending and receiving sections of the beam stations are to be capable of working simultaneously, the aerial system of the sending station being so designed as to concentrate the emitted waves within an angle of 30 degrees, and the receiving section to have a similar aerial system, designed to focus the received waves.

The stations are to be capable of communication at a speed of 100 five-letter words per minute each way (exclusive of repetitions) during a daily average of 18 hours between England and Canada, 11 hours between England and South Africa, 12 hours between England and India, and 7 hours between England and Australia. Upon the completion of the stations, the Company is to give a demonstration by actual working for seven consecutive days that they fulfil these conditions, and if the Engineer-in-Chief of the Post Office is satisfied with the results of the demonstration, the stations are to be handed over to the Postmaster-General, and one half of the agreed purchase price is to be paid to the Company.

After six months' working to the satisfaction of the Engineer-in-Chief, a further 25% of the price is to be paid, and the remaining 25% is to be paid at the end of a further period of six months, if the stations have continued to work to the satisfaction of the Engineer-in-chief. If the stations do not satisfactorily comply with the tests prescribed at any of these stages, the Postmaster-General is free to reject the stations, and the Company must, in that eventuality, refund any money which has already been paid in respect of them.

The Company undertakes that any telegrams for this country which come under the control of their affiliated Companies in the Dominions shall be forwarded to a Government station in this country. A similar undertaking is given by the Company as regards any telegrams for the Continent of Europe which are not ordered by the senders for transmission by some other route. At the outset the rates of charge to the public are not to exceed two-thirds of the corresponding cable rates in force at the date of the Agreement (July 1924), except in the case of Canada, where the rates are not to exceed the existing cable rates.

There are ten steel masts at each station 287 feet high, with a cross arm at the top, 90 feet broad, placed in two lines of five each, each line being at right angles to the direction of communication. The aerial and reflector system consists of a number of vertical wires suspended from the triatics attached to the cross-arms, the aerial wires being in front of the masts and the reflector wires behind. Each wire is brought down to a point within a few feet of the ground, and balance weights are provided to minimise the movement of the wires. Each vertical wire is connected to the transmitter or receiver as the case may be by a copper tube supported about two feet above the ground, and so arranged that there is the same length of tubing from the transmitter or receiver to each vertical wire.

The stations at Bodmin and Bridgwater for communication with Canada and South Africa are practically completed and are expected to be ready for the official demonstration within the next few weeks. The stations near Grimsby and Skegness for communication with India and Australia are expected to be ready for preliminary tests about the middle of November. All the stations are to be controlled from the Central Telegraph Office, London, which will thus be placed in direct communication with Montreal, Cape Town, Bombay and Sydney."

The physical size of these stations was massive, with their masts clearly visible for miles.

View of the Tetney Beam Wireless Station,
1928, showing the large masts used for the service.

It seems that the UK were somewhat behind their counterparts in providing Beam Wireless services, and in July 1923 the press took the then unusual step of criticising the Postmaster-General (Sir Laming Worthington-Evans) for this delay:

"It is said that the new Postmaster-General has so far made no move in regard to the Empire wireless chain. His predecessor, Sir William Joynson-Hicks, took up the question with great energy and had a scheme complete, including a settlement with the Marconi Company - which means with Australia, South Africa, and Canada - but Sir Worthington-Evans has apparently not moved in the matter. This country is far behind other great nations in wireless and the Empire is left helpless. But perhaps we shall get a move on shortly - at least we will all hope so."

It appears that the criticism served its purpose, as Sir Worthington-Evans left his role some 6 months later.

Despite this, politics seemed to be responsible for the delay in introducing the 'Beam Wireless' system in the United Kingdom. A detailed report from August 1924 sheds light on the issues which needed to be overcome:

"Mr. Hartshorn, in the House of Commons yesterday, moved a resolution approving the contract made between the Marconi Wireless Telegraph Company and the Postmaster-General for the construction of a wireless telegraph station on the beam system in connection with the Imperial wireless scheme. He pointed out that in the past the Dominions and ourselves had developed different systems of wireless, and eventually a committee was set up which recommended a wireless system now embodied in the agreement.

Canada had definitely decided on arrangements for erection of a station on the beam system, and he understood that a similar arrangement had been made for Australia and South Africa. Col. Moore Brabazon expressed approval of the policy of the Government that connection with Empire wireless stations should be controlled by the State, but the Marconi Company had manoeuvred us into a very difficult position, because they had control in all cases at the other end, and we had to communicate through them.

Sir H. Brittain supported the agreement, which he said marked a definite step forward in the development of the most important of Empire communications - an Empire wireless chain.

Baker said that, having regard to the history of the negotiations between the Marconi Company and the Government, the House should look very carefully and critically at this proposal. He was given to understand that the beam system was immature, and that it had not been successfully demonstrated to the Government. He asked that before the contract was signed, Senator Marconi should be called to demonstrate the process of the beam experiment before the Wireless Telegraphic Commission.

Without this safeguard he should have the gravest fear as to the future. He could not understand why the beam system had been introduced into their system at the present time. Our experience as a nation appeared to be entirely different in relation to this problem from that of the French. He gathered that the high power station outside Paris was able to communicate with the whole of the French Colonies, and that the French Government was entirely satisfied with that system. When the possibilities of the high power station were so well known and appreciated, it was difficult to understand why at that point the Marconi Company should endeavour to persuade the Government to embark on something which was untried and unknown.

Everything which would produce delay had been resorted to by that Company in the past, and public propaganda had been designed to prove that the Post Office as a Government Department was unfit and unsuited to undertake this work, and that the nation would be well advised to permit the Marconi Company to perform the service for them.

He thought they should have definite specification of the stations to be erected in order that they might be satisfied that a fair price was being paid for the stations. He was inclined to think that the event of the Postmaster-General not being satisfied with the stations, and declaring that they did not meet the terms of the memorandum, the Marconi Company would find themselves in the favourable position of having the stations erected and under their control, and with public opinion inflamed at the delay which had taken place it would be able successfully to demand a licence which should enable it to operate this particular system between this country and the Dominions.

Mr. Hartshorn, replying to questions, said the agreement provided for a year's experimental working of the system. It had been signed and could not now be amended. The motion was agreed."

However, Marconi himself had been conducting experiments of his own from his station at Poldhu, Cornwall, which in July 1924 had succeeded in communicating directly with Australia on a wavelength of 92 metres. Communication was established over a path of darkness, which led him to believe that high-speed signalling would be extremely suitable for this path, clearing as much traffic during the darkness hours as the long-wave stations would clear in 24 hours.

Use of directional aerials utilising a reflecting screen were involved, providing greater efficiency and economy.

It is worth considering the construction of these massive 'Beam' aerials, which, with minor local differences, were consistent throughout the worldwide network.

Ten masts were erected at Bodmin and Bridgwater, five being used for the service to Canada, and five for South Africa, two bays being utilised for each wavelength.

At Grimsby and Skegness, eight masts were constructed, as only one wavelength was provided for communication with Australia.

Each row of masts was arranged in a line at right angles to the great circle bearing of the distant station. The masts were of steel-lattice construction with an overall height of 287 feet, except in the case of the Australian masts which were 260 feet. The cross arms were 90 feet long and also of steel-lattice construction.

Each mast had four legs and cross bracing of angle section, the vertical members being made up of 5" by 5" by ¾" angles. The legs were attached by holding down bolts to concrete blocks, each 8 feet square. The centres of the concrete blocks were 12 feet apart and the masts were 12 feet square throughout their length.

The masts were spaced at 650 feet and each was provided with one set of 4 stays attached at 216 feet above the ground. The end masts also had 4 back stays to balance the strain of the triatics.

The aerial and reflector wires consisted of a number of vertical wires of No. 16 SWG copper-clad steel, suspended from triatics which were attached to the cross arms, the aerial wires being in front of the masts and the reflector wires behind in the direction of propagation.

The triatics were suspended on the double catenary suspension system, and tails of varying length were inserted between the triatics and the aerial and reflector wires to allow for sag in the triatics and to ensure that the average height of the wires followed the contour of the ground surface.

Each vertical wire was brought down to a point within a few feet from the ground, and balance weights were provided to ensure an even tension on each individual wire and to minimise the movement of wires during high winds.

Inductances were inserted in the aerial wires, and the reflector wires were broken up into insulated sections, the number of wires, their distance apart and the number of inductances being dependent on the wavelength employed.

At the associated transmitting stations, precise alignment of the masts was of extreme importance, and the construction of them involved a great deal of care, using fixes on the sun and stars to ensure they were square-on to the shortest Great Circle path to the desired destination.

The angle of elevation was between 10 and 15 degrees from the horizontal for ranges of over 2000 miles. Phasing coils were used between the half-wave sections to bring the aerials into phase, although these were later replaced by a 'zig-zag' array to produce non-radiating phase reversing.

The concentric feeder to the aerials consisted of air-insulated concentric copper tubes, held apart by porcelain spacers, in effect an early version of the modern co-axial cable. The feeder was carried on iron supports driven deep into the ground, with a distribution unit keeping the supply to all the aerial wires in phase.

The transmitters used at these transmitting stations had a valve drive unit taking less than 100 Watts, with the output being amplified by three successive stages. Duplicate drives were installed where different wavelengths were used for daytime and night transmissions, which allowed a change of wavelength in around 10 minutes.

New valves were designed to cope with the high frequencies being used, and were known as the oil-cooled 'CAT' (Cooled Anode Transmitting) valves.

Returning to the receiving stations, the following were the particulars of the Canadian short-wave bays, similarly replicated at other stations in the network:

There were 24 aerial wires spaced at 19' 0" apart and twice the number of reflector wires; the latter were spaced at 9' 6" and staggered in respect to the aerials so that the projection of an aerial wire on the reflector plane was midway between two wires.

The distance between the line of aerial and reflector wires was 40 feet, or slightly greater than ¾ of a wavelength. In the case of the longer waves, the distance was reduced to about ¼ wavelength. Each vertical aerial was divided into three straight sections separated by phasing inductances.

These lengths were unequal and varied from ⅝ of a wavelength at the top to about 1¾ wavelengths from the bottom inductance to the auto-transformer.

The reflector wires were divided into 8 insulated sections, each approximately ½ wavelength in length.

The Australian beam system of reflectors was somewhat different, as in that case the wires were in one uninterrupted length earthed at the bottom.

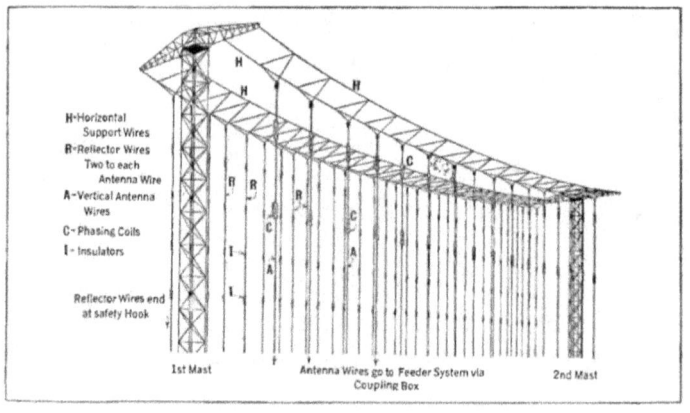

The construction of the 'Beam' aerial.

Due to the success of these tests, the original plan for high-power long-wave stations was cancelled, and a new plan using these short-wave 'Beam' stations was organised.

Within the United Kingdom, a network of stations was established; a transmitter site at Bodmin, Cornwall, working to Canada and South Africa was built, with a double receiving station at Bridgwater, Somerset. Another transmitting station at Dorchester was erected, for communication with New York and South America, with a receiving station at Somerton, Somerset.

A third station with transmitters at Grimsby with a double receiving station at Skegness providing communication with India and Australia was also established, again operated by the G.P.O. and controlled from a Central Radio Office in London.

Each destination country would have similar stations and aerials erected to allow for two-way communication with the UK stations.

Compared with the cost of the Rugby long-wave station, it became clear that these new stations made economic sense; The Rugby station was estimated to have cost £350,000, whilst the Bodmin/Bridgwater station worked out at approximately £55,000 per transmitter and receiver; about one-sixth of the cost, which would have been even lower if omni-directional transmissions were used instead of the directional beams.

Elsewhere in the world, other links were being established, such as Paris-Djibouti, Madagascar-South America, Germany-South America and Netherlands-Indonesia. The German transmitter at Nauen operated three transmitters, 'POR13' operating on 18 metres, 'POW' operating on 28 metres with 50 kW, and 'AGA' on 26 metres, with 10 kW.

Further testing in December 1924 on a 30 metre wavelength established communication with Montreal, Rio de Janeiro, Buenos Aires and Sydney during daylight, but these paths were not so reliable during darkness hours.

The national press were naturally excited about the possibility of global communication, and an article from February 1925 conveyed their sense of wonderment:

"As already announced, Beam Wireless stations for communicating with Canada are to be erected in the West. A receiving station is to be at North Petherton, near Bridgwater, and the sending station in Cornwall. By the agreement with the Marconi Company, communication with Canada is to be established within 26 weeks from the date on which the sites are placed at their disposal, and it is expected this will be done by the end of the month.

The site of the Cornwall station is at Horras Farm, near Lanivet, 2½ miles S.W. of Bodmin.

When Sir William Mitchell Thompson, Postmaster-General recently announced in Parliament that the sites for the Canadian and South African wireless beam stations, near Bodmin and near Bridgwater, had been selected, he said it was hoped that the negotiations for the purchase of the particular properties would be completed in time to place the work in the hands of the contractors by the end of the present month.

The site selected at North Petherton is Copse Farm, comprising 203 acres, with four adjoining fields, and the carrying out of the scheme will be one of much interest to Bridgwater and the district. It appears that the first site at Bodmin will be for the transmitting apparatus, and the second (North Petherton) for the receiving end of the station, which is to be constructed by the Marconi Company and to be ready for work within six months from the date of the provision of the site by the Post Office. This station is intended to be part of the scheme for ensuring wireless communication with the Dominions, and it is provided that similar stations shall eventually be erected in India, Canada, Australia, and South Africa.

At present the Canadian station is the most forward; some parts of the work are almost completed and the masts have been built. There are two such stations in Canada, one at Montreal to work with England, and the other at Vancouver to work with Australia. There will ultimately be four stations on this side to communicate with each of the Dominions named. The South African station is under construction near Cape Town; the Australian Company has accepted the Marconi Company's tender for the construction of a station, but the site has not yet been fixed; and India has also agreed on the terms of its contract; so that the preliminary contracts for the Empire Beam Wireless stations are now complete.

The Marconi Company is also at work on the construction of its own Beam Wireless stations for communicating with America and the Continent, at a site two miles from Dorchester. In each case the wavelength is below 100 metres. The height of the aerial mast and the length of the aerial are dependent on the wavelength, and the masts are about 300 feet and about 650 feet apart.

The Beam Wireless is understood to be proving more effective and cheaper than the broadcast. The wave is concentrated directionally upon the station for which it is intended, and the result is not only greater reliability for the message, but also considerable economy in transmission."

The wonderfully-named *'Jack Broadcaster'* (probably not his real name) reported in his *'Wireless Notes'* column that:

"It is over two years ago since I predicted in these columns that the day was not far distant when all long distance wireless telegraphy would be made by the Marconi 'Beam' system, of which so much has been heard these last few days.

For some time past messages have been successfully despatched at the rate of more than 100 words a minute in both directions simultaneously, for 11 hours out of the 24. The great advantage of Beam Wireless is that it can be worked on an unusually low power, and therefore, at a much cheaper cost. The beam system, of course, is directional, and instead of sending a message which may be picked up all over the world, it will only send it in one direction. It needs only about 20 kilowatts power to work a powerful beam station, whereas 700 to 800 kilowatts are needed for a station such as the Rugby superpower station, which broadcasts throughout the world. The experts are of the opinion that this is the first great step to long distance wireless telephony being placed on a commercial footing. Already a man sitting in London has spoken to Australia, but results are as yet too uncertain to make it of practical use. With the beam system, however, it brings the possibilities much nearer."

The *'Western Daily Press'* in its edition of 22nd October 1926, reported the new 'Imperial Wireless System' in glowing and patriotic tones:

"The dream of Imperial Wireless is about to be realised, and it is a coincidence that the point of realisation should be reached during the session of the Imperial Conference in London. It is announced that, as from next Sunday midnight, the new 'beam' system of transmission between England and Canada will be inaugurated, and this will form the first section of a complete Imperial system. A trial was made yesterday, when a body of British journalists dispatched a message to Canada, and received an answer in a very short time. The experiment appears to have been a complete success.

The details of the system are highly technical, but the general public will learn with interest, and even astonishment, that a new system of inter-transmission by wireless signals has been perfected to the extent that it will be capable of far exceeding the highest speed yet attained by means of the present system of wireless.

It is claimed that as a result of the tests carried out between Canada and Great Britain that by the 'beam' system 1,250 letters per minute in each direction can be transmitted - thus making a total of 2,500 letters per minute - is a remarkable record of efficiency, and, in association therewith, it is furthermore claimed that there can be no interference with the messages that are passing to and fro, thus eliminating delays which are very often vexatious, and even dangerous. These claims will of course, have to be substantiated by further experimentation, but the experts are sanguine as to the results.

As far as can be ascertained at present, the general public will not stand to reap any immediate advantage from the installation of the new system, because the charges of the new system will be the same as the cable rates, and the Post Office will, no doubt, take care that the cables enjoy a certain amount of protection, or such of them, at any rate, come within the aegis of Post Office supervision.

It is claimed that the initial cost of erecting a 'beam' station is far less than that of an ordinary wireless station, and that the upkeep charges will be comparatively moderate. As Mr. Marconi has explained, the beam system is by far the most speedy method of intercommunication yet devised, and the speed is at present limited only by the mechanical limitations of the transmitting and recording instruments. In spite of some sceptical criticisms, the prospects of the beam appear to be rosy, and further developments in the clarity and speed of transmission will be watched with interest.

One of the stations has been erected near Bridgwater, and it will commence operations next Sunday as one of the pioneers of a real all-Imperial wireless system. Other stations will be set up in various parts of the country. Some of these are already completed, so that the inauguration of the new system will be made under auspicious conditions. No one would be safe to prophesise the extent to which this new departure will be developed, because wireless is a comparatively new device, and no doubt, capable of enormous improvement. At all events, up to the present the sceptics have been confuted as were those of an earlier day, who cast doubts on the potentialities of the railway train. Almost all new discoveries in the realm of science are held suspect by those whose imaginations are lacking in vividness, and whose hopes are measured by the restricted horizon of their own mentality.

But the vision of Empire as we see it now is wide, and we must be prepared to give free play to all the potentialities that may lurk in it. But the laying down of the British Empire under a closely-woven network of wireless seems to be heading rapidly towards the *fait accompli*. Preparations have been made for the speedy inclusion of South Africa, Australia, and India within the orbit of the 'beam' system. It is computed that within the course of a few months these preparations will be complete, so that the whole Empire will be welded together by a fresh and apparently reliable, though invisible, link. It is a marvellous achievement towards the shrinkage of time and space - a process that has been going for generations. During the two past reigns and the current one, the rate of progress has been phenomenal, and the records rank high on the pages of history. The value of the wireless network to the Imperial system is well-nigh beyond calculation. Strategically, socially, and commercially, it will provide new guarantees, new opportunities, and new and binding ties between the Motherland and the Dominions overseas whose delegates are today gathered as guests under the 'old roof-tree'.

The same newspaper (and probably the same author) continued to wax eloquently about the new service in its issue of 30th December 1926 under the heading of 'The Latest Marvel':

"Though bristling with technicalities that may be understood but hazily by the average man, the popular imagination will be stirred by the advent of the latest marvel for the shrinkage of space and time. It is nothing less than the bridging of vast areas by wireless telephony respecting which we published some particulars yesterday. The system will be inaugurated by the establishment of intercommunication between England and America, under the direct supervision of the Post Office, whose appointed experts will apply the essential tests.

It is not expected that there will be any failure to pass these tests, because experimental work has been in progress for a considerable time past, and the experience gained in the course of this has been turned to the very best account.

Members of the younger generation, who, accustomed to aeroplanes, motor-cars, submarines, and all the rest, are too prone to regard even the most startling innovations commonplace, will accept the new and far-reaching system of radio-telephony with equanimity. But to those who lived before wireless in any form existed, the present development will be regarded as trenching on the miraculous.

That a man sitting in his office, or in his home, in London should be able to carry on a *'viva voce'* conversation with another man in New York is a proposition that might have been relegated to the category of fairy tales; but it is an achievement which will be common enough in the immediate future.

Individuals in two vast and teeming cities thousands of miles apart will be brought into touch with each other. They will transact their business in a few moments, perhaps, shut down their instruments, and give the time that has been saved to the accomplishment of other tasks. At first the fee will be high, and not unnaturally so, but in course of time it is believed there will be substantial abatements, even though the capital outlay has been enormous.

The minimum charge for a conversation lasting three minutes between New York and London will be fifteen pounds, which works out at little over penny per mile over the stretch of three thousand miles. Above the minimum there is to be a graduated scale of fees, and it may pay the keen business man moments of emergency to pay the price.

The radio-telephonic system between Great Britain and the United States about to be opened will be in the nature of a pioneer, and it is expected that soon there will be a wide extension of it, and this will synchronise with the immense development of the 'beam' system of wireless, by means of which all the Dominions Overseas will be linked up with the Mother Country and with each other. In this country five 'beam' stations have been fixed, and that for the reception of messages from Canada and South Africa is Bridgwater.

It has been provisionally arranged that the official inaugural greetings over the radio telephonic system will be exchanged between King George from Buckingham Palace and President Coolidge from the White House, Washington, the representative capitals of the Anglo-Saxon race. Extravagant ideas as to the capabilities of the new radio-telephonic system may be entertained, but what has already been done would seem to afford some guarantee that there will be little risk of failure. It is already obvious that these cheaper and ultra-modern methods of long-distance intercommunication must have their influence on the scale of cable rates.

The position is the cable versus the 'beam' transmission and the radio-telephone, and it is expected in well-informed quarters that the rivalry between the two groups will become intense. Competition, however, is the life of business, and the general public nearly always stands to gain when the element of competition begins to operate.

Just as the 'beam' system of wireless marked a great advance on the systems that had preceded it, the system of radio-telephony will mark an equally definite advance on the 'beam', which is capable of transmitting 2,500 letters per minute over the completed circuit. The speed of the telephonic messages will have to be regulated by amateur senders and receivers thereof. One person may transmit in the course of one minute much more than another could transmit in double that time.

The users of the radio-telephonic system will be at liberty to regulate the pace according to their own inclinations and skill; but the great thing is that it renders the transmission of the spoken voice a certainty, long as atmospherics can be barred.

The Post Office, however, has made arrangements for allowing special remissions in the charges when 'atmospherics' occur. It should not be hastily concluded that the radio-telephonic message is projected through the air from one terminal to another. Some part of the way at each end is covered by the land lines. A message from London, for instance, passes over the special underground long distance lines to the wireless station whence it is dispatched on its way across the Atlantic, and then carried along a stretch of American land lines. With messages from America the process is reversed. Though to the layman the process may seem to be complicated, it is none the less extremely ingenious and interesting, and the radio-telephonic system marks another milestone along one of the great highways of scientific progress."

It seemed the country was becoming obsessed with the new 'Beam Wireless' system, and the erection of new masts at the Portishead Transmitting Station (the long-range maritime service) caused great excitement in the *'Western Daily Press'* on 18th October 1927, when it erroneously reported that:

"A great deal of curiosity has been aroused by the erection of huge masts on a high point of the coast at Portishead. This is one of the new Marconi type of 'Beam' stations to provide high-speed wireless service on the short-wave beam system with the principal Dominions. In the West of England there are three such stations. The first to be established was the one near Bridgwater, and there is another at Bodmin, Cornwall. The latter is the transmitting station used for communication with Canada and South Africa, and the Bridgwater station the receiving station for these services.

It is said that the masts at Portishead are 300 feet high, and as the ground upon which they stand is probably 300 feet above sea level, the masts can be seen from over a wide area. The view, we are assured, from the top of the masts is a remarkable one on a clear day. Exeter Cathedral can be easily recognised. At Bridgwater, the five masts are erected in a straight line at right angles to the direction in which communication is to be established. These masts are 277 feet high, each having a cross arm at the top measuring 90 feet from end to end."

Of course, the aerial masts at the Portishead site were designed for the long-range maritime radio service (Portishead Radio), and it would appear that the author of the article assumed that these new masts were simply a further extension of the Beam Wireless service.

The Beam Wireless Stations were regarded as one of the wonders of the world during the late 1920s and early 1930s, where the novelty of instant communication to the far corners of the world was at its peak. The wireless pioneers of Marconi, Franklin, and many others became celebrities in their own right.

Marconi died on 20th July 1937 and as a tribute all wireless stations around the world observed a 2-minute silence the following day. In the United Kingdom all radio transmissions fell silent, as reported below:

"At six o'clock to-night, the hour fixed for the funeral, all B.B.C. transmitters and all the Post Office wireless telegraph and wireless telephone stations in the British Isles will close down for two minutes.

The Post Office has sent the following message to all ships: "As a token to the late Marchese Marconi, all stations are asked to stop transmitting between 17.00 and 17.02 G.M.T., July 21, except in cases of extreme urgency."

The Postmaster General's announcement of this tribute to Marconi adds: "Rugby, the most powerful wireless station in the world, which links the British Empire by radio telephone, will pay this mark of respect as well as the Post Office coast wireless stations which maintain communication with ships on the seven seas."

Franklin, who has been regarded as the 'father of the Beam Wireless system', retired in 1939, and both national and local press in the United Kingdom were quick to pay tribute to his work:

"Mr. C. S. Franklin, the man who designed most of the early 2LO broadcasting station, has never spoken into the microphone at Broadcasting House, except for testing purposes. Regretfully the man who perfected the system directional wireless telegraphy known as the Marconi Beam System, told of his retirement from the Marconi Company, enforced by attaining his 60th birthday.

During his 40 years with the Marconi Company, Mr. Franklin worked with the late Senatore Marconi both in this country and Italy, when the early experiments with short waves were being conducted.

He is proud of the fact that he was on board the Cunarder 'Philadelphia' in 1902 when Marconi was making his tests between Poldhu, Cornwall, and Newfoundland, to gauge the signals across the Atlantic. A granite column at Poldhu, near the site of the wireless station from which Marconi heard the famous signal at Newfoundland in 1901, bears the name of Mr. Franklin and commemorates the fact that he was one of the pioneers. Having given the world cheap wireless transmission by his discoveries of the beam system, one of his last efforts before retirement was the designing of the television aerial at Alexandra Palace.

The concentric feeder system for supplying energy radio aerials and enabling extremely high frequency currents transmitted along line without loss, for which he was responsible, are essential to television transmission circuits. In May 1938, he was presented with the James Alfred Ewing Medal for his contributions to the science of engineering in the field of research. Mr. Franklin was born at Walthamstow, but it is his intention to build a bungalow at Poldhu, the 'home' of Beam Wireless, and to spend his retirement working harder than ever in experiments.

"I am out of a job now", he said ruefully, "so I have to fill in my time somehow."

He died on 10th December 1964, aged 81.

CHAPTER 2

THE UNITED KINGDOM BEAM
WIRELESS NETWORK

Once the UK network had been formally established,
the 'Empiradio' (as the service became known) services
comprised:

Transmitting Station	Call Sign	Wavelength in Metres	Receiving Station	Date Service Opened
Bodmin (England)	GBK	16.574 32.397	Yamachiche (Canada)	25th October 1926
Bodmin (England)	GBJ	16.146 34.013	Milnerton (South Africa)	5th July 1927
Grimsby (England)	GBH	25.906	Rockbank (Australia)	8th April 1927
Grimsby (England)	GBI	16.216 34.168	Dhond (India)	6th September 1927
Drummondville (Canada)	CGA	16.501 32.128	Bridgwater (England)	25th October 1926
Drummondville (Canada)	CFA	24.793	Rockbank (Australia)	16th June 1928
Klipheuvel (South Africa)	VNB	16.007 33.708	Bridgwater (England)	5th July 1927
Ballan (Australia)	VIZ	25.728	Skegness (England)	8th April 1927
Ballan (Australia)	VYZ	24.958 16.286	Yamachiche (Canada)	16th June 1928
Kirkee (India)	VWZ	34.483	Skegness (England)	6th September 1927

The overseas stations are outside of the remit of this
book, but brief details of the United Kingdom stations are
included to put the Somerset stations into context. It is

recommended that Paul Hawkins' excellent book about the Point-to-Point wireless service (entitled 'Point-to-Point') is referred to for further in-depth reading about the UK network.

Christmas traffic in 1927 averaged 140,000 words per day, and in 1928, 180,000 words per day. A dip of traffic during the summer was noted, but it was clear that average traffic figures were rising year by year.

As an example, when the Imperial Cable network broke down in June 1927, traffic on the Canadian and Australian Beam Services doubled, rising to 50,000 words per day. When the South African service became operational later that month, this figure was maintained and increased in September 1927 when the Indian station opened. Traffic figures over the four duplex circuits averaged around 90,000 words per day.

Competitors of the Marconi Company had been watching the development of the Beam system with great interest (but original scepticism), but soon realised that the Beam principle was an efficient and effective system. However, these competitors were some years behind the Marconi Company in development, which gave Marconi a distinct advantage.

There was no system anywhere else in the world which could work at such as high speed as the Marconi Beam system, and it was noted in the *'Marconi Review'* of November 1928 that "it is interesting to note that our competitors, in order to obtain signals of the requisite strength and dependability for commercial service, are now increasing the area of their aerials to approach nearer to the standard which the Marconi Company established when the Imperial Beam stations were erected."

One system, known as the 'Projector System' was designed by American and French engineers, which was described in the same publication:

"Designs of projector aerials, rectangular and lozenge-shaped networks have been employed, which are broken up by insulators so as to constitute several groups of half-wave aerials in series, suspended one above the other, and these groups are fed either from one end or from the middle. A cheaper feeder system is in consequence obtained, but the character of the beam is again adversely affected.

The Franklin method of feeding the aerial does not introduce this trouble; if the frequency alters, the intensity of the Beam is affected, but its direction and concentration do not change, so that the cost of the feeder system is fully justified."

With regard to operational requirements, it is important to understand that traffic must have been sent as soon as it was tendered to the station, meaning that the load on each station would fluctuate rapidly, and that the equipment at each station must have been capable of handling such loads, especially at busy times.

Politics was never far away from the development of the Beam Wireless Service, and the network was discussed in Parliament on many occasions. One fascinating exchange took place on 21st May 1928 when the matter of the ownership and authority of the stations was discussed in great depth:

A speech was made by Otho Nicholson, the then MP for Abbey, in the House of Commons:

"The hon. Member for East Bristol (Mr. W. Baker) who opened the debate, stated that there were six separate authorities dealing with the telegraphic and telephonic communications of the British Empire. I believe that there are really eight, and there is no single authority responsible for the co-ordination and the development of these various services to make them of the greatest commercial, strategic and political advantage to the Empire.

The need for some such authority is absolute. First of all, because Empire unity depends on rapid and efficient communication between the home Government and the Dominion Governments and between the United Kingdom and our Dominions. Secondly, because there are other countries and particularly the United States of America, who are developing telephonic and telegraphic communications to such an extent that I believe they are a danger to this country if we wish to maintain control of these communications. And, thirdly, because so rapid and extensive have been these developments that the future outlook is so unstabilised that we shall be left behind unless we have unified control. We must have some unified control that is strong enough and courageous enough, should the circumstances arise to scrap the old system and introduce the new. Empire unity demands prompt and efficient telegraphic service. I believe that that is an accepted fact.

We cannot regard the Dominions as vast lands thinly populated merely separated from us by thousands of miles of sea. Whether they like it or not, they are drawn into world politics, and the home Government cannot ignore them and must consult with them in all questions of international politics. To do so with promptitude and dispatch, they must have the latest and most efficient form of communication. This applies more particularly to this country than to any other country owing to the great distances which separate the home Government and the Governments of our Colonies. Therefore, I say there is not only great need for a single controlling authority but a very pressing one.

As regards the question of other countries developing wireless telegraphic communications, the most active of which is the United States, recently concessions have been obtained for a wireless service between the Argentine and Spain by an American company, and again an American company proposes to establish a service between the Pacific and Japan and the Far East. That same company has purchased the Sayville Wireless Station, which until lately belonged to the American Government, in order to set up communication with European countries.

The American company has also obtained control of the telegraph communications between America and Spain, and I believe I am right in saying that American financiers to a very large extent control the radio and telegraph companies of Germany. These developments have the approval and the support of the American Government. The American Government have set up a Federal Radio Commission for the purpose of allotting short wavelengths to American companies, to the naval and military authorities, and to others who are interested in America. Already, a very large number of these short wavelengths have been allotted, and the beam system is a system which is worked on comparatively short wavelengths.

There are not more than 500 or 600 wavelengths available for the whole of the world communication and one American company has already had allotted to it, has applied for, and states that it requires at least 225 of these wavelengths. The deduction to be drawn from these facts is that we have to be very careful indeed to see that we get our proper proportion of those wavelengths.

In view of the importance of telegraphic communications to the Empire, it ought not to be left to the individual action of eight different authorities to obtain and maintain an efficient service. This responsibility should be in the hands of a single controlling organisation of the whole Empire. In the minds of the past generation, the submarine cable was considered to be the last word in scientific invention. Only three years ago we thought that the large high-powered wireless stations were equally the last word in scientific invention, but we made a mistake.

Hardly before these stations had become stabilised they were superseded by this new system known as the beam, and the beam system, I believe, will be used entirely in the future for all communications of any distance from this country. The beam system has already passed the experimental stage. It is in operation as far as the Empire is concerned between this country and Canada, South Africa, Australia and India, and these four beam circuits are carrying 30,000,000 words per year, and are capable of carrying five times that amount.

The beam system has many advantages over the cables and the long-wave wireless stations. First of all, let us take the question of costs. Every hon. Member knows what a costly thing it is to lay down a submarine cable. We all know what an enormous amount of money the large Post Office station at Rugby costs the taxpayers of this country - a matter of £500,000. On the other hand, these beam stations can be erected for a matter of £100,000. As regards the speed, the cable is only capable of transmitting messages at a rate of approximately 45 words a minute.

The large long-wave wireless stations are also limited in the speed with which they can transmit messages, probably somewhere between 20 and 30 words a minute, largely owing to the enormous current with which they have to deal. The beam system, which is dealing with a comparatively small amount of energy, is able to transmit messages at the rate of 200 words per minute, and under the most adverse conditions can keep up an average speed of 100 words per minute for the whole 24 hours. I understand that very shortly these beam stations will be capable of increasing their speed to probably something like 600 words a minute.

One right hon. Member has already given the House the rates compared with the cables and wireless. In three cases, the rates of the Beam Wireless are 4d. a word less than those of the cable. It is only in regard to a Canadian service where the two rates remain the same. With regard to the question of efficiency - cables versus wireless - cables sometimes develop faults, and, when they do, it is an extremely costly business to send a ship out, first of all, to drag for the cable, and then to mend it.

The normal sort of breakdown which one gets in a wireless station is one which is comparatively easy to repair - probably a burnt-out valve. If it is something larger, such as a burnt-out armature, there is probably a spare one which can quickly be put into commission. The question of secrecy is one which is always held up on behalf of the cables as opposed to wireless, and it is one with which I do not agree. We must remember that these beam stations can transmit at the present moment, as I have already said, at the rate of 200 words a minute. That in itself makes it an extremely difficult thing for anybody to intercept. There is a new invention known, I believe, as the cryptograph which automatically codes and decodes any message and is capable of altering that code every sentence - if you like, every word; if you like still more, every alternate letter. That is going to make it practically impossible for any person to decode these messages.

There is the question of the narrowness of the beam. Unless you happen to be in the path of the beam, it is very nearly impossible to intercept these messages. May I give the House an illustration of what I mean? A little time ago messages were being sent by the Beam Wireless stations from South America to London, as the focus point. Messages were received in London strong enough to be automatically recorded, but those same messages were only audible on earphones to the German station which was listening for them. They were not strong enough to work the automatic machinery. With a slight adjustment of the aerial in order to widen the beam those messages were capable of being automatically recorded both in London and Berlin. No doubt in future it will be possible considerably to narrow the beam and, by doing so, increase the secrecy.

As regards the future development of the beam, I believe that if the stations were in the hands of private enterprise, we should find that they would develop very much more quickly than they will in the hands of the State. It is possible at the present time to turn these beam stations into wireless telephony stations, and at the same time as you are sending messages on the Morse code it is possible to superimpose the human voice on the same wave. We have the same thing going on at the Post Office Rugby station, and they charge us £5 a minute for communication with America.

The ASSISTANT POSTMASTER-GENERAL (Viscount Wolmer): The charge is now £3.

Mr. NICHOLSON: The Assistant Postmaster-General informs me that the charge has been reduced to £3. It is time that it was reduced still further. It is possible to introduce telephony to beam stations and to work it at a handsome profit of £1 per minute. There is a new invention which I hope will be introduced into the beam system - an invention for the transmitting of photographs. Perhaps that does not appeal to hon. Members unless I explain that messages themselves can be photographed and facsimile messages can be sent over the wireless. In a comparatively short time we may be reading that the full 24 page issue of the *'Daily Mail'*, complete with picture page at the end, will be published every day simultaneously in every capital of our Dominions.

Mr. DUNCAN: What about the *'Daily Herald'*?

Mr. NICHOLSON: The *'Daily Herald'* will have the same possibilities if only the Government will give facilities to their beam stations. For the purpose of development, and of watching the development of experimenting, particularly, one unified control authority is essential, on account of:

(1) The importance of Imperial communications;

(2) The competition for world control of wireless telegraphy communications, and;

(3) The rapid development and instability of the wireless telegraph companies, at any rate, at the present time.

The question which arises is whether the Post Office should be that body, and I ask myself, does the Post Office record in relation to inland telegraphs inspire confidence? One has only to read the Hardman Lever Committee's Report where it refers to: The atmosphere of inertia and the lack of resiliency of the telegraph service, and the unsuitability of Civil Service conditions to apply to a business undertaking, to realise that, in their opinion, the Post Office are not a fit and proper body to undertake this work.

Does the Post Office inspire confidence with regard to wireless telegraphy? I would refer the House to the history of their Rugby Station - a monument of lack of foresight on the part of the Post Office engineers. That station was put up for the express purpose, as I understood it, of communication with our Colonies at any time when we wished to do so.

What do we find this station doing at the present time? Whenever I have listened to it, I have never heard it doing anything other than sending messages to ships at sea, most of them Press messages. The service which at one time it had established with Cairo has now been taken over by the beam station. Not long ago, the Post Office told us that the only way to create efficient wireless telegraph communication with the Empire was by a series of short steps, or relay stations.

I do not believe that the Post Office has the necessary vision. One has only to refer to the question of the Pacific cable. The Pacific Cable Board is composed entirely of the representatives of Governments. They laid down a cable which cost £2,750,000, and every Government agreed to it except the Government of Canada. Not one of them except perhaps the Government of Canada realised the possibilities of the beam system.

One hon. Member in this House, during the debate on the Pacific Cable Board, referred to the Pacific cable as a very valuable asset. I think the right hon. Member for Seaham (Mr. Webb) referred to it as "the most successful social enterprise that could possibly be imagined," and yet today we are taking steps to consider what we can do to save this successful Socialistic enterprise from private enterprise competition.

Mr. AMMON: Surely, the hon. Member is mistaken. It is from Post Office competition or public competition that we are seeking to save it. The wireless system belongs to the Post Office.

Mr. NICHOLSON: I think I am right in saying that the Post Office are not the only people who own beam stations from which competition may come. I think the Marconi Company have stations, and I think the Americans are setting up short-wave stations. The Secretary of the Post Office Workers' Union, speaking the other day at Weston-super-Mare, said that the beam service was the most astounding verdict in favour of State control as against private control. I think that is a most astounding claim. Who experimented with the beam? Was it the Government? No. It was a private individual. Who made the beam a practical service? Was it the State? No. It was a private company.

Who installed the beam stations? Was it a Government Department? No. It was a private company. Who put these stations in order? Again, it was a private company, and not the Government.

The State did not take any interest in the matter until it was conclusively proved that the stations were capable of carrying on an efficient service. It was then that they took them over. The beam telegraph system is a triumph of private enterprise. Finally, as against Government control and interference, I would point out that every other country has deliberately allowed the beam system to remain in the hands of private enterprise. Why should this country be handicapped by allowing these systems of communication to be taken over by the State?"

It was clear that the UK press continued to be extremely excited at the developments in the service, and in February 1929 reports appeared lauding 'Marconi's Victory over Space':

"Successful experiments which were carried out with the Marconi system during the weekend have made the time appreciably closer when the transmission and reception of telegraphic messages in facsimile instead of by the Morse system will become the ordinary procedure. It will soon be possible to flash a picture of oneself with a message in one's own handwriting to friends thousands of miles across the ocean as simply and easily as a cablegram is sent today.

The stock phrase, 'Even a child can do it', so often applied to the operation of scientific inventions nowadays is particularly applicable to the latest development of the new Marconi facsimile telegraph system, for a little child has done it, sending his picture across the Atlantic by wireless to his grandparents at home as easily as if was posting a New Year's card to them from the next street.

"It happened like this," said a Marconi official describing the incident. "We have recently been conducting a number of facsimile tests between New York and Somerton, and in the course of these tests little Bryan Davis, the son of the English engineer in charge of the American end of the experiments, was able to send a picture of himself in his new hat with his New Year message to his grandparents at Chelmsford.

During the weekend a number of pictures and hand-written messages have been flashed across the Atlantic at high speeds on the short wave beam circuit in operation between New York and Somerton, and an important step thus taken towards the solution of the problem of rapid facsimile transmission.

The system is designed to replace the Morse working on busy circuits, and make it possible for Marconigrams to be received in the actual handwriting of the sender. When such a system is universally adopted, it is expected that it will reduce the telegraph operating costs, with the result that facsimile telegrams, instead of costing more than ordinary telegrams, should cost less.

Marconi's system differs greatly, both mechanically and electrically, from any other, and, while there are certain difficulties still be overcome before practical perfection is attained, is now able to transmit two images, each eight inches by ten inches, in less than 20 minutes. These messages have the advantage of unimpeachable accuracy and the possibility of including facsimile matter of all kinds in the telegrams transmitted - a particularly valuable feature when complicated columns of figures and diagrams are required to be sent.

This new system is in every way suitable for the transmission of pictures over land lines well as wireless, the speed of transmission of pictures depending on the quality of the land lines.

Yesterday's tests included the sending of cartoons and views of American buildings from Rocky Point, America, to England. They were received almost simultaneously in a dark, tiny hut at Somerton, Somerset, the speed of the transmission across the Atlantic being 186,000 miles a second.

The receiving apparatus looks like an extraordinary elaborate type of cylindrical gramophone. A piece of sensitised paper is placed on a cylinder, and a spot of light, a veritable optical paint brush, revolves from inside the cylinder on to the paper. The cylinder and paper move sideways one hundredth part of an inch every revolution of the light until the whole of the paper has been covered by the light spot. The paper is then developed and in less than a minute is visible in England, complete in every detail."

The UK was very much at the forefront of the worldwide network, although numerous international circuits became operational throughout the 1920s and 1930s, many both to and from the USA where large transmitting and receiving stations were constructed for this purpose.

Politics have never been far away from commerce and industry, and the ownership and regulation of the Beam Wireless Stations was no different. In May 1928 the merger of the Beam and Cable services was discussed in Parliament:

"On the motion for the adjournment of the Commons, W. J. Baker (Labour, Bristol E.) raised a discussion on the question of the Imperial and Cable Wireless Services. He said that the Prime Minister said it was necessary of the interests of national security that there should be a wireless station in this country capable of communicating with the Dominions and owned and operated by the State.

In March this year, however, following the assembling of the Imperial Wireless and Cable Conference in London in January last, an announcement was made that it was proposed to transfer the Post Office cable and beam stations to private enterprise. He referred to a financial merger between the Marconi Wireless Company and the Eastern Telegraph Cable Company, and said he understood that that merger was a calculated attempt to force the hands of the Government.

They hoped to take over the Post Office wireless and cable system and control the independent wireless and cable companies in the Dominions. The fact was that the Marconi Company knew how great were the potential profits of the Government Beam Service, and that the Post Office Beam Service, although in its infancy, was already showing a very handsome profit. The cable companies were frankly afraid of the success of the Post Office Beam Service, and they had entered into a financial merger as the only way of escaping from that competition.

He objected to our Imperial wireless communications being handed two groups, one of which had a record for scandalous mismanagement. "We should not allow any Minister or civil servant to enter the service of any company or combine with which he previously had official negotiations" *(hear, hear)* and he asked for the names and rank of all Government servants who had transferred themselves to the Marconi or the Cable Companies.

If it were the fact that a decision had been reached on this matter by the Imperial Wireless Conference, then the Government had adopted a course that was fatal to the national interest in connection with services that were so necessary to our national well-being, and the chief result of the conference had been to enrich share manipulators and similar people. He understood that the parties to the merger were acting as if the transfer was a *fait accompli*. He hoped that was not the case.

Sir Hamar Greenwood (Conservative, Walthamstow E.) disagreed with Mr Baker's conclusions in favour of State control, which, in view of lack of capital, made for under-development. The great experiment with State-owned cables and the beam system had been excellent but the time had come when it should be transferred to private enterprise.

'There seems to be an atmosphere of corruption and wire-pulling about wireless communication and the cable companies without parallel almost in the history of any commercial moment' said Mr. Ammon (Labour, Camberwell N.). The Government were threatened that if they did not come to heel the American companies and the British merger would wage war against British industries to get complete control.

Major Hills (Conservative, Ripon) urged that coordination was more advantageous than competition, which, if unrestricted, would destroy the cables.

Sir J. Gilmour, Secretary for Scotland, Chairman of the Imperial Conference that has been inquiring into the situation that has arisen as a result of the competition between the wireless and cable services, said there had been thirty meetings of the conference since its first meeting in January last, and the conference was still in being.

The problem before the conference was not an easy one. The merger between the Marconi Company and the Eastern Telegraph Company, for which the Government had responsibility, was made subject to satisfactory arrangements being made with the Governments of Great Britain and the Dominions and India. The possible reactions of the merger had had to be examined in great detail by the conference. Conversations and communications were still proceeding, and he was unable to say at present when they might be brought to conclusion.

It should be appreciated that the conference was only empowered to make recommendations to the various Governments represented. It could not itself come to any operative conclusion. That was a matter for the Governments themselves. Before any definite action was taken, the matter would be brought before Parliament. The only desire of the conference was to come to a conclusion as soon as they could with due regard to the difficulties of the situation they had to deal with. *(Hear, hear)*.

Mr. Hartshorn (Labour, Ogmore) expressed that nothing was to be done in this matter until it had been brought before Parliament. The motion was by leave withdrawn."

The Post Office workers also weighed in by protesting against the transfer as reported on 28th June 1928:

"A resolution protesting against the rumoured transfer of the Beam Wireless and Cable Services from Government ownership to private enterprise was passed by a meeting of Post Office workers in London last night.

The meeting was organised by the London District Council of the Union of Post Office Workers and Mr A. C. Winyard, who presided, said the insidious attacks made on State ownership and now concentrated on the Beam Wireless and the cable services were taking the same line as Italy.

The interest behind that move would not stop until they had grasped the whole of the postal services which could called remunerative. Mr. J. W. Bowen, secretary of the Union, said the Post Office had shown that State control was entirely successful. An opportunity should be given to the House of Commons and the public of deciding whether any serious change was to take place. The Post Office had got private enterprise beaten to a frazzle. Many of the private cable companies had had a fat time and were now cross over Post Office competition.

The British Post Office had shown vision and experimented with wireless before Marconi came to this country. There was no cause for the transfer of the services to improved private enterprise."

What must be recorded, though, is the understanding of long-range radio propagation which research throughout the 1920s had enabled many radio engineers and scientists to understand. In fact, it is due to such research which enabled the development of radio communication throughout the 20th century.

The August 1929 edition of *'Modern Wireless'* carried an article entitled 'Lazy Waves', with the strapline "Beam Radio has advanced so rapidly that we are apt to forget that a special wireless technique and new theories had to be developed before the present degree of excellence could be attained. The way 'beam waves' behave is interestingly described in this article."

It is well worth reproducing this article in full, as it concisely describes the methodology and understanding of radio propagation at the time, and is indeed still valid today.

"There is always something interesting in the *'Marconi Review'*, and, in a recent issue, Mr. T. L. Eckersley, the brother of Captain Eckersley, has an extremely interesting article on Beam Wireless Waves.

Mr. Eckersley proves pretty well conclusively that short waves, similar to those used by the beam system, are quite lackadaisical compared, for instance, with the waves used by the B.B.C. Furthermore, Beam Wireless waves do, to a certain extent, wander off the path and, besides lingering on their journey, behave in other ways which provide interesting scientific problems.

In some recent experiments which Mr. Eckersley conducted, he set up a special direction finder at Chelmsford. The idea was to check the bearings of the various transmissions from Marconi Beam Stations. As a result, these short-wave transmissions were found to behave in more than one peculiar way. For example, transmissions from the Grimsby Beam Station, which devotes most of its time to sending to Australia on a wavelength of 26 metres, is in a direction slightly west by north from the Marconi research station at Chelmsford. Curiously enough, however, the direction finder showed it to be bearing west-north-west during the morning and east-south-east in the afternoon!

More or less similar discrepancies were discovered to exist when this particular experiment was applied to other Marconi Beam Stations at Bodmin and Dorchester, and it was further found out that these beam waves took a longer time to reach their distance than theoretical calculations indicated. In fact, it appeared that ·08 of a second was wasted by these waves *en route*. In other words, they dawdled on the way to an extent which, although in our artificial system of time may be regarded as negligible, had, nevertheless, important significance.

Mr. Eckersley offers a very interesting explanation in the *'Marconi Review'*, and, as one might expect, his explanation brings in the problem of the Heaviside layer, which, as my readers know, exists above the earth's upper atmosphere. According to the article, it has been discovered that the apparent position of the transmitting beam station, as indicated by the direction-finding apparatus, is the position where the transmitting waves strike the Heaviside layer and are scattered; while the waves reaching the receiving station actually come from a point on the Heaviside layer sometimes thousands of miles away from the original source.

Now, this sort of thing is what complicates the theory of short-wave transmission and reception, and Mr. Eckersley is to be congratulated on being the first investigator to point out this scattering effect. It is generally thought to be the cause of various echo phenomena, an example of which was recently pointed out in *'Modern Wireless'* in connection with the Oslo station.

There was a general idea once that when a beam wave strikes the Heaviside layer it is reflected chiefly in one specific direction, though, of course, in a very minor degree in all directions, for the most perfect beam apparatus in the world has not yet enabled engineers to direct a wireless beam of energy without some slight diffusion of this energy. As a result, waves arrive at the receiver from more than one direction; and as some of the short waves regard distance as almost non-existent, some of them probably travel round the world in opposite directions, with the result that the same signal reaches its destination by various routes and, consequently, at different times.

For example, if a dot in the Morse code is transmitted to one of the beam stations, the scattering effect results in the signal arriving by various routes. This can prove very annoying in the case of a Morse dot, because the difference in time would result in the dot being lengthened as a signal, and perhaps to inexperienced ears sounding something like a dash. More particularly, this becomes extremely aggravating in high-speed reception work, as blurring results and, consequently, confusion in signal reception."

The understanding of such phenomena was applied to the Beam Stations in order to improve efficiency and quality; equipment provided by Marconi and other manufacturers was updated accordingly and the aerials themselves modified if required."

The 28th March 1930 was an important day in the history of the Beam Wireless Service when a long and detailed debate took place in Parliament regarding the future of the service and the involvement of other companies with the Post Office in the service. Some countries took the opportunity of seeking guarantees from the Post Office, such as Australia:

"The Australian Post Office is demanding a guarantee that the proposed wireless telephone service from Rugby shall be available at certain definite hours before accepting the British Government's plan. The Post Office is also asking that the fee for a three minutes' conversation shall be £6 instead of £9 as at first proposed, and that of this sum the British Post Office shall take £3, the Australian Post Office 15s., and Amalgamated Wireless, which will operate the service, £2, 5s."

The debate became heated at times, but suffice it to say that a great deal of rationalisation was planned which saw some of the original beam stations in the United Kingdom closed by the end of the decade.

However, technical developments and understanding of radio propagation continued to improve, with radiotelephone services becoming available via the network, as reported in June 1932:

"The opening of the Post Office's Beam Wireless telephony service to Canada within the next few weeks will be a triumph for the methods which have been throughout adopted by the engineers of the Department. Arrangements for the India and Egypt services are also well advanced, and it is expected that these will be opened in the autumn. The criticisms of Post Office methods of construction which were freely made two years ago have been proved by events to be groundless.

While many other wireless concerns have suffered serious inconvenience from the approach of the worst period of the sunspot cycle, the Post Office services have been little affected. Sunspots affect the electrically-charged layers of the upper atmosphere, which reflect wireless waves round the earth, and a steady increase is at present taking place in the lengths of short waves which can be successfully used.

The Post Office plan, since adopted by the Radio Corporation of America and the Spanish-South American service, was to build short aerials, which could be comparatively cheaply replaced, and to employ a large number of alternative wavelengths in order to meet the demands for telephone service without delays.

It is pointed out that the arguments for such low aerials which were put forward in the House of Commons by Mr. Lees-Smith, the then Postmaster-General, have been entirely substantiated. Quoting from an expert report, Mr. Lees-Smith said; "There appears be no reason for supposing that an effective aerial array would need masts of over 180 feet in height, and there is no doubt that the costs of masts increases very rapidly when they are over 800 feet in height."

The Marconi Company were pioneers in beam development and the Marconi beam stations already erected are handicapped somewhat by this fact. The Post Office system of aerial arrays being carried on lower masts could, therefore, probably made more effective than the Marconi array at lower capital cost.

Reference was also made in the report to the greater ease with which variety of wavelengths could be used to meet changing conditions at different hours of the day. It is this adaptability which has saved the situation now that the 11-year sunspot cycle has become a serious factor in commercial transmission.

"The American service is now being satisfactorily operated on the four wavelengths which we already have," stated a Post Office representative. If conditions become bad it will probably be necessary to add a fifth and higher wavelength, but this can done at relatively small expense."

Publicity for the service continued to progress and indeed took numerous forms, with advertisements and posters appearing in Post Offices around the country, and even in stamp booklets as pictured below:

The service publicised in stamp booklets of the time.

A few examples of early advertising material.

Staffing of the stations initially proved difficult as the traffic handled by the network was much greater than had been expected or even provided for. Inexperienced staff had to be employed to assist in handling the increased traffic, but as they became more 'expert' in the work, recruitment was reduced, and gradually staffing and traffic figures balanced out.

*The Indian and Australian receivers, Skegness.
Photograph dated 1928.*

With regard to the equipment at each station, most stations were similarly equipped, with the aerials 'beamed' to the associated transmit/receive station. Each station was fitted with the following basic apparatus:

1. The Aerial and reflector.

2. The Aerial feeder system.

3. The transmitter.

4. The receiver.

A high-level description of each component part can be shown as:

1. **Aerial and reflector (Bodmin used as an example):** There are two sets of five masts, one set supporting the two aerials of the Canadian circuit, and the other set supporting

the two aerials of the South African circuit. The layout of each row of five masts is arranged so that the Great Circle bearing on the distant station with which that particular Beam aerial is designed to communicate, is at right angles to the line of the masts.

The masts are 287 ft. high with 90 ft. cross arms, and are spaced 650 ft. apart.

Each aerial occupies two bays between the masts, and consists of a parallel sheet of elements made up of a number of vertical doublets linked by phasing coils. The aerial wires are spaced about one quarter wavelength from a screen of twice as many reflector wires.

The aerial arrangement is such that currents fed into the parallel wires of the aerial are all in phase. Under this condition the energy radiated from the individual wires cancels out in the plane of the wires but adds in the direction at right angles to this plane.

The effect of the reflector is to cut off the back radiation from the aerial and strengthen it in front, and the total result is a strong beam of radiation confined almost entirely to one direction only and spread over an angle determined by the dimensions of the aerial.

The aerial systems of the transmitting and receiving stations are identical.

2. The masts and aerial systems at Grimsby and Skegness employ a single reflector with an aerial on each side of it, so that by changing over from one aerial to the

other the direction of the beam projected round the earth can be altered.

An evocative illustration of the GPO Radio-Telephone Services.

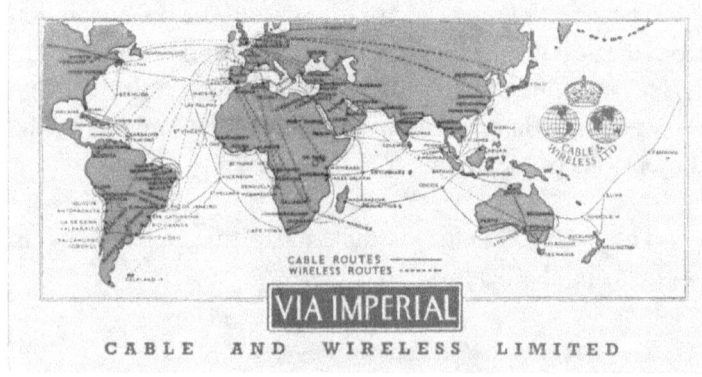

A Cable and Wireless illustration of their network.

3. **Feeder system:** Current is fed into the aerial from the transmitter or conveyed from the aerial to the receiver by means of a special feeder system of concentric copper tubes, the outer one of which is earthed, and so arranged that the length of feeder to each individual aerial element is electrically the same, which ensures that the currents in all the aerial wires are in phase.

4. **The Transmitter (Bodmin station as an example):** The Bodmin station is equipped with two transmitters, with the four-panel Canadian transmitter on the left and the South African transmitter on the right. The station precision wavemeters are on a bench in the centre. The rectifiers and the smoothing circuits for the H.T. D.C. supply to the valves are in another part of the building.

Stability of the wavelength is obtained in the first instance by a control circuit, and in the second place by the extreme care taken in the construction of the set. The drive, or master oscillator, which maintains the intermediate and main oscillatory circuits on its own frequency, is carefully screened from the other circuits except at the point where it is weakly coupled to the next stage known as No. 3 Magnifier. This is in part an amplifying circuit, and in part a stabilising circuit which by acting as a buffer between the drive and power circuits proper, helps to maintain the constancy of the drive wavelength when keying.

As in the case of the drive, the No. 3 Magnifier is carefully screened except at the point where it couples to the next stage known as No. 2 Magnifier, which in turn is coupled to the grid circuits of No. 1 Magnifier, the main power oscillator of the transmitter in which oil-cooled valves are employed.

The four panels of the transmitter comprise:

No. 1 Panel: The No 1 Magnifier.

No. 2 Panel: The No 2 and No 3 Magnifiers and the drive for the first optional wavelength.

No. 3 Panel: The No 2 and No 3 Magnifiers and the drive for the second optional wavelength.

No. 4 Panel: The main and sub-absorbing and keying circuits, which, by means of two oil-cooled valves in parallel, divert the H.T. supply through resistances during the spacing periods, and so keep a constant load on the generators.

When changing over from the day to the night wave therefore, Panel No. 1 is re-adjusted, Panel No. 2 is cut out, Panel No. 3 is switched in, and Panel No. 4 continues as before.

5. **The Marconi Beam Receiver:** In addition to the switchboard, there are nine boxed units, the function of which are as follows:

- Feeder Terminal Unit.

- Intermediate Tuning Unit and First Heterodyne Unit.

- H.F. Amplifier and Filter Unit.

- First Rectifier.

- First L.F. Amplifier and Filter Unit.

- Second L.F. Amplifier and Filter Unit.

- Second Rectifier.

- Limiting Unit.

- Modulating Unit.

The feeder connection is made at the Feeder Terminal Unit, which according to the wavelength received may either be a simple tuning circuit, or a tuned valve circuit to magnify the input signals.

The Intermediate Tuning Unit and First Heterodyne Unit includes an intermediate tuning circuit and first stage heterodyne which produces a beat wave of about 1,600 m. The signal then enters the H.F. Amplifier and Filter Unit, which includes a 3-stage amplifier and filter which gives uniform amplification over a frequency band of 5,000 cycles.

The First Rectifier converts the signals to a wavelength of 10,000 metres, which are passed to a second amplifying filtering unit for the 10,000 m. signal, limiting the band width to 3,000 cycles. There are two further amplifying and filtering stages, and then the Second Rectifier comprises the final push-pull rectifiers and second 'listening' circuit.

The Limiting Unit contains a modulating valve and circuits to allow the HF oscillations generated by the first heterodyne to be modulated at 1,200 cycles, so that the operator can listen in at the first stage of the receiver.

Of course, the stations were subject to ongoing radio propagation conditions prevalent at the time, and operating frequencies were changed dependent on the time of the day and the darkness path to and from the destination. As an indication of this, the following wavelengths were utilised:

To Canada:	16.574 metres and 32.397 metres.
To South Africa:	16.146 metres and 34.013 metres.
To India:	16.216 metres and 34.168 metres.
To Australia:	25.906 metres.
From Canada:	16.501 metres and 32.128 metres.
From South Africa:	16.077 metres and 33.708 metres.
From India:	16.286 metres and 34.483 metres.
From Australia:	25.728 metres.

It will be noted that Australia has no alternative wavelength; this is due to the fact that radio propagation to and from that country can take place either on a Short Path or Long Path, the wavelength selected being the optimum predicted for reliable communication."

Wireless and cable rates (per word) at the time were:

Between UK and	Cable Rates Jan 1927	Cable Rates Jan 1928	Beam Wireless Rates
Australia	2s 6d	2s 0d	1s 8d
South Africa	2s 0d	1s 8d	1s 4d
India	1s 8d	1s 5d	1s 1d
United States	9d	9d	9d

Traffic figures for the time indicate just how busy the stations were. The monthly average of the Canadian service was 5,000,000 words in September-November 1927, and 6,000,000 words in March-May 1928. The averages for the Australian service during the same periods were 8,000,000 and 9,000,000 words respectively. The South African service remained consistent at around 9,000,000 words. The service to India, which opened on September 6th 1927 reported a monthly average of 9,000,000 words between October-November, and the subsequent average between March-May 1928 was 10,500,000 words.

In the year endjng December 6th 1928, the British Government's share of the gross revenue from the four Beam Services was £440,000, and of the net revenue, £166,000 before deducting depreciation and interest.

September 29th 1929 saw the transfer of the Post Office Beam Wireless Network to a new company, as advised in a national press release:

"The Postmaster-General announces that on Sunday, the working of the Imperial cables and of the Beam Wireless services to the Dominions and India (i.e., the telegraph system known as 'Imperial' and 'Empiradio') will be transferred from the Post Office to Imperial and International Communications (Ltd.).

This is the company recently formed to take over - as recommended by the Imperial Wireless and Cable Conference of 1928 - the Imperial and Empiradio services of the Post Office, as well as the services of the Pacific Cable Board, the Eastern Telegraph Company (Ltd.), and

Marconi's Wireless Telegraph Company (Ltd.) As From September 30, telegrams for the Imperial and Empiradio services will be accepted at any office of the Eastern Telegraph Company (Ltd.) or of Marconi's Wireless Telegraph Company (Ltd.).

The Imperial Cables and Empiradio services remain under British control and the Postmaster-General expresses the hope that the public support gained while the services have been owned and operated by the Post Office will continue to be given to them after their transfer to Imperial and International Communication (Ltd.). All classes of telegrams to be sent *'Via Imperial'* or *'Via Empiradio'*, as well as by other routes, may still, after the transfer of operation, be handed in at any postal telegraph office at existing rates of charge."

The Skegness receiving station, 1928.

The name of the new holding company would be Cables & Wireless Ltd., although this would be slightly changed to the singular Cable & Wireless Ltd. in 1934.

A Parliamentary debate on the subject of the Beam Wireless network and the Overseas Wireless Telephony service took place in great depth in Parliament on 26th March 1930. It is fascinating to follow the debate in detail as it gives an overview into how the Government of the time viewed the service, and explains the rationale in centralising the overseas wireless services at Rugby and Baldock. However, the content is too detailed and long-winded to reproduce in full; suffice it to say that opinions were divided on this matter.

This of course had a great effect with regard to the future of the receiving stations at Bridgwater and Somerton, as well as the associated transmitting stations, as we shall soon discover.

An article published in *'Wireless World'* in December 1932, and written by the former deputy Inspector of Wireless Telegraphy, Lt. CDr. Chetwode Crawley M.I.E.E., gives a snapshot of the global network at that time.

"The economic situation has prevented any striking developments in long-range wireless telegraph services, and, indeed, few of even the present circuits can have been working at full capacity. The station of the League of Nations was perhaps the most interesting development in this category during the year. This station, or, rather, group of stations, was completed in February, and within an hour of opening was in use for important communications with

Japan. The transmitting station is at Prangins, about 15 miles from the receiving station, which is in a suburb of Geneva.

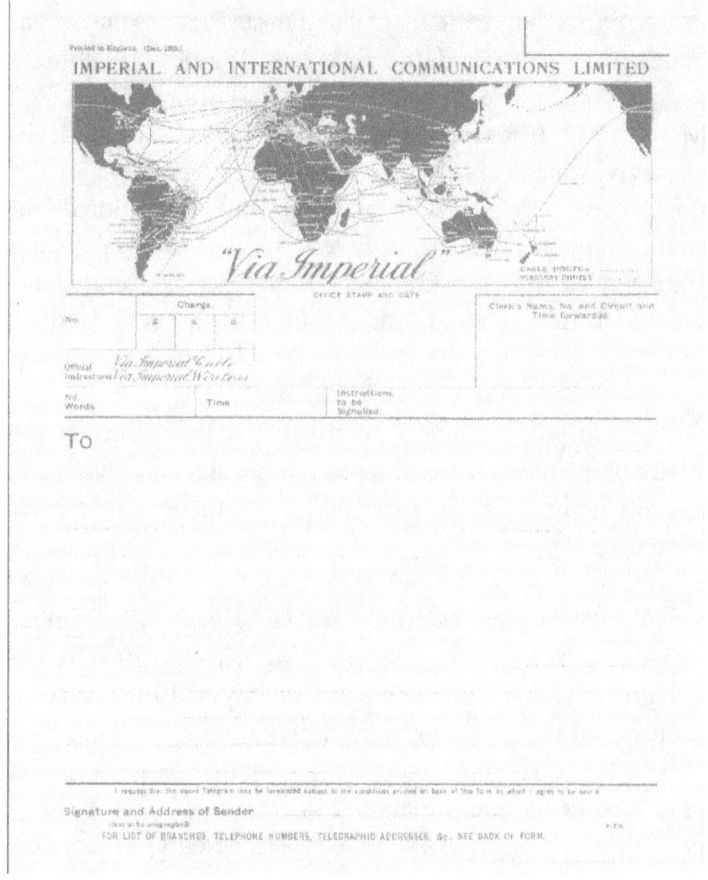

A blank Imperial telegram form, used for delivery of international messages.

It was quite an international affair. The buildings and some of the masts and machinery were of Swiss manufacture, and the wireless apparatus was supplied by British, French, German, and Dutch companies. Provision is made for beam transmission and reception on short waves, and all-round communication on long waves, so that the station is able to communicate with all parts of the world. A telephone circuit between this station and Japan has just been established.

Another service of interest was opened in May, when a new Imperial wireless telegraph link was established between this country and the Central African territories, by the inauguration of a service with a beam station at Salisbury, in Southern Rhodesia.

Wireless telephony, being of such recent growth compared with telegraphy, has had a better chance of expansion, even under the present adverse conditions, and its development for communication over great distances has progressed satisfactorily throughout the year, though, of course, much of this development has consisted in the extension of wireless channels by land lines and cables. At the end of 1931 this country was connected telephonically with the Berlin-Siam wireless circuit, and on January 1st this year with the Berlin-Venezuela circuit.

On January 1st, too, the wireless telephone circuit between England and New Zealand was extended by line to practically the whole of Europe. In February the England-South Africa wireless service was opened, and in March the transatlantic service was extended to Bermuda and the Sandwich Islands, the former being a wireless extension from New York, the latter a land-line extension from New York to San Francisco, and thence by wireless to the Sandwich Islands.

In April, the England-Australia service was extended by line to Perth in Western Australia, and in May the England-South Africa service was extended by line to Europe. A wireless service was opened in June between this country and Egypt (Cairo and Alexandria), and was extended by line to Port Said in October.

In July the direct England-Canada service was opened. Up till then the service with Canada had been made by way of New York. A month later the England-South Africa service was extended from the Cape Province by line to Johannesburg and Pretoria, and later to other cities. In September it was extended by wireless to transatlantic liners.

In December most of the wireless telephone services were extended by line to Lisbon. The following is a summary of the long-range wireless telephone circuits now in operation from this country, and most of them can be connected together in London, which may be looked on as the switching centre of the world:

London-New York (four channels): Connecting Europe with the United States, Canada, Cuba, Mexico, Bermuda, and the Sandwich Islands;

London-Sydney (one channel): Connecting with Australia and New Zealand;

London-Cape Town (one channel): Connecting with South Africa;

London-Buenos Aires (one channel): Connecting with Argentina, Uruguay, and Chile;

London-Rio de Janeiro (one channel): Connecting with Brazil;

London-Montreal (one channel): Connecting with Canada direct;

London-Cairo (one channel): Connecting with Egypt.

There are also other services available by line to foreign countries, and thence by wireless, e.g., Java via Amsterdam, French Indo-China via Paris, Siam and Venezuela via Berlin."

1936 saw a fascinating experiment using the Beam Wireless circuits, reported nationally on 13th February of that year:

"Sitting a chair in his study at the Royal Institution in London, Sir William Bragg, the famous scientist, by the simple act of lighting a candle, flooded with light the New York Museum of Science and Industry in the Rockefeller Centre, more than 3.000 miles away.

The occasion was the opening of this new science centre. In the presence of a distinguished gathering, which included Dr. Albert Einstein.

Sir William sat down before the candle at the Royal Institution at 3.35 this morning (10.35 p.m. in New York) and before he performed the lighting ceremony he explained to his distant audience that he was sitting before the table of Faraday, the pioneer of electricity and magnetism.

They heard Sir William strike a match, which applied to the candle, and in an instant the entrance hall of the museum in New York was filled with brilliant light from two mercury vapour lamps. The energy of the small candle flame in London had been converted into ethereal radiations which were sent across the Atlantic over the Post Office short wave radio circuit. When the radiation reached New York they automatically switched on a Westinghouse lamp, which was manufactured 50 years ago.

The light of this lamp thereupon operated a photo-electric cell, which, in turn, brought into play a number of electric switches. The lamps controlled by the switches completed the circuit by means of an electrical current, which produced the mercury vapour, and within an instant the striking of match in London the darkness gave way to brilliant illumination in the New York Museum.

Afterwards Sir William Bragg delivered a brief wireless address to his fallow scientists, which was continued by Dr. Einstein and the Mayor of York the museum, as well as by Miss Amelia Earhart, the airwoman, and Dr. Robert A. Millikan, who were in California.

Experiments in the transmission of wireless energy were first undertaken by the Marchese Marconi, who in 1930 from his yacht *'Elettra'*, anchored off using transmitting apparatus no bigger than a small wardrobe, switched on the lights of an electrical and radio exhibition in Sydney (Australia). The impulse which carried this experiment to success travelled via the Beam Wireless station of Imperial and International Communications at Somerton in

Somerset, and reached Sydney after an aerial journey of at least 14.000 miles.

Four years later the new Orient liner *'Orion'*, which was being built at Barrow-in-Furness was launched by wireless from Australia by the Duke of Gloucester, who was attending the Victoria centenary celebrations. The Duke sent the vessel down the slipway by the simple act of pressing a button in Brisbane."

The global point-to-point radio service continued in one form or another for many years, especially to and from countries where reliable communication links had not yet been established or were deemed to be unreliable. However, it was clear that the system remained at the mercy of prevailing radio propagation or atmospheric conditions, as a report from 27th January 1938 notes:

"The display of aurora borealis, northern lights, on Tuesday caused considerable disturbance in the Post Office radiotelephone services. Since last Friday, it is stated, phenomena of a similar nature have had a disturbing effect on the short wave telephone circuits, and on Tuesday night most these circuits became unworkable early in the evening, and remained so throughout the night.

It was possible to maintain the service with North America by means of the Rugby long wave circuit. All the Post Office radiotelephone services were in operation yesterday with the exception of the short wave circuits to the United States and Canada. It not yet known whether the peak of the disturbances has been passed."

A somewhat disturbing discovery was published in 1938 in that the transmission of bacteria via Beam Wireless had been proven:

"In a London laboratory the other week, a group of scientists tried to find out more about bacteria in the interests of health, made the terrifying discovery that it is possible to transmit bacteria through space by Beam Wireless.

Little imagination is required to visualize what this would mean in time of war.

Gas-masks and even hermetically-sealed underground shelters would be useless against a death-ray attack by the enemy. The wireless beam would penetrate through any obstruction just as we now receive broadcasting with indoor aerials.

The significance of this newest scientific discovery is fully realised - which explains why identities are closely guarded by the authorities."

No further developments appear to have been reported.

In the late 1939, Lambert & Butler issued a set of cigarette cards entitled 'Interesting Sidelights on the work of the G.P.O., two of which depicted the work of the Central Telegraph Office (C.T.O.).

The rear of the cards give a brief description of the equipment: The 'Picture Telegraphy' card is described as: "Pictures, designs or documents may be telegraphed in facsimile between London and many European towns. The picture is fixed round a revolving drum and a spot of light passes over it. The light reflected from the picture actuates a photo-cell which sets up electrical impulses proportionate to the tones of the picture. These impulses are transmitted after amplification over a telephone circuit to the receiving apparatus, where a light, varying in intensity with the impulses, plays on a film or piece of sensitized paper on a similar drum. The film or paper is then developed like an ordinary photograph"

Of course, as regards the handling of 'picture telegraphy' over the point-to-point network, the same principle would apply, the phrase 'telephone circuit' being replace by 'radio circuit'.

The second card was described as: "The teleprinter is now the standard machine used for the transmission of inland telegrams. In appearance it is rather like a large typewriter. The operator types the message on the keyboard and the letters are reproduced in typewritten characters on a paper tape associated with a similar machine at the distant end of the circuit. The paper tape which contains the message is gummed upon a telegram form and sent out for delivery. A single teleprinter can be used simultaneously for the transmission and receipt of telegrams, and when so used a total output of over 200 messages an hour can be reached. We show a teleprinter operator of the Central Telegraph Office at work."

Similarly, such machines would have been used at each of the radio circuit in the point-to-point network, the transmission path being over radio links rather than land lines.

PICTURE TELEGRAPHY
C.T.O. SENDING APPARATUS

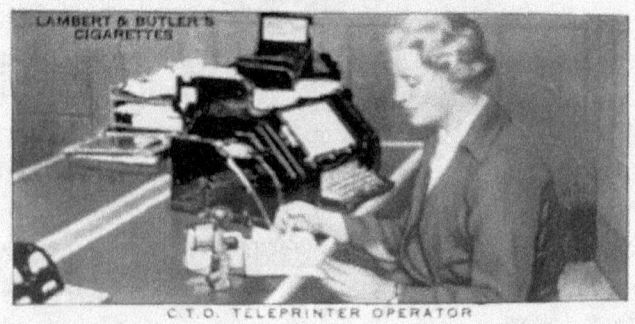

C.T.O. TELEPRINTER OPERATOR

Cigarette cards depicting the work of the C.T.O.

1939/1940 saw the closure of the Bridgwater/Bodmin and Grimsby/Skegness sites, with their services being taken over by other sites in the United Kingdom, with a greater reliance on the Rugby transmitter site and Baldock receiving station.

The service expanded during the 1930s and other sites were added to the ones listed above; transmitting sites at Dorchester and Ongar were established, and receiving sites at Somerton and Baldock complemented these.

However, the decision to centralise the services in 1938 caused the closure of the Bridgwater/Bodmin and Tetney/ Skegness stations. The closure of the Skegness station in particular was sadly reported on 8th May 1940 in the *'Skegness Standard'*:

"The 'passing' of the Beam Wireless Station at Skegness is regretted in many quarters, the personnel having been popular in a number of circles, while the material loss to property owners, tradespeople and others which is represented by the removal of sixteen members of the technical and outside staffs will be very considerable.

As most local people are well aware, the removal of the station is in no way connected with the war. The scheme for greater centralisation in this country of the services which the Company performs has been under way for two or three years past. Technical reasons, linked with more economical working have been the sole consideration, and the transfer would have been made even had the war not occurred.

Members of the staff took the formal farewell of their 'chief', Mr. E. E. Frankis, at a little presentation gathering the other day. The whole of the personnel were present, including the night staff, who had foregone their sleep for the purpose.

The gift of the technical staff was a silver tray, suitably inscribed, and that of the outside staff a leather wallet case. These were handed over to the accompaniment of suitable little speeches, and it was obvious that the recipient was greatly touched by the tangible expressions of goodwill.

Some of the staff will remain under Mr. Frankis at the new station, and the remainder dispersed among other stations controlled by the company. A member of the staff told a 'Standard' representative that was extremely sorry that the transfer had come about.

The Skegness station, he said, had been very happy one indeed, and he and his colleagues looked upon Skegness as one the nicest towns at which they had been located, as servants of the company. As some of them had been at Skegness much longer than at any other station, and had been comfortable here, they would leave the place and its people with real regret, coupled with best wishes for the continued growth and prosperity of the resort."

Wartime brought extra usage of the stations, and many were camouflaged (some better than others) to reduce the possibility of aerial attack; in the event, very few stations (if any) saw damage by enemy action and they continued to operate effectively during the conflict.

The company was very keen to promote their continued operation throughout the conflict, with a selection of patriotic advertisements in the national and local press:

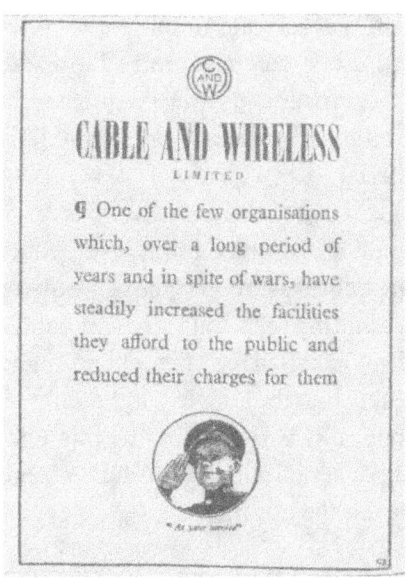

Wartime advertisements from 1944.

The text of the right-hand advertisement features the phrase 'Messenger of the Free Peoples', and goes on to say in patriotic style:

"Britain is now the European nerve centre of the Allied Nations. The governments of Occupied Europe are in Britain and the Empire. The military headquarters of the forces of the new world are 'over here.' From Britain radiate the messages that control this mighty effort. Approximately two million words pass through the central telegraph station of Cable and Wireless every day.

The heavy strain which has been thrown upon Cable and Wireless resources is being met and overcome by the skill and loyalty of the staff. No difficulties to be met in the future can prove insuperable to men who do not I know the word 'defeat.'"

After the war, use of the services returned to fully commercial operation, and users were encouraged to make the most of their return; advertisements were published in the national press with offers of reduced rates, amongst standard promotional material.

The advantages of radio communication over cable communication was clear; cables were very susceptible to damage through lifting or cutting, especially where cables were located in countries occupied by the enemy. The use of radio precluded such hostile activity, although the lack of secrecy would naturally be an issue. To this end, various means of encoding or scrambling radio signals were developed and applied during the conflict.

On the cessation of hostilities in 1945, control of the stations reverted back to Cable & Wireless, although the company was nationalised in 1947, with assets integrated with those of the Post Office.

Half-price pictures being offered and customers encouraged to contact American G.I. friends from 1946 and 1947.

In 1950, the control of the overseas services of Cable & Wireless Ltd. from the United Kingdom was transferred to the Post Office as part of the nationalisation scheme introduced in 1947. At the same time, the radio beam stations leased to Cable & Wireless were returned to the Post Office, in effect reversing the 1929 merger referred to earlier in this publication.

The service was heavily advertised, and numerous leaflets and posters were produced to entice potential users to avail themselves of the facility. Although the posters were keen to use the generic term 'cable', many transmissions of messages actually took place via radio, using the beam system.

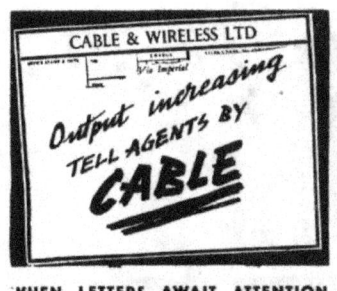

Similar Cable and Wireless advertisements from the post-war period

In 1954, senders of greetings telegrams using the radio service were advised to tender their messages well in advance, as an announcement by the Post Office on 21st December of that year stated:

"Because of the possibilities of radio difficulties, the Post Office urgently advises the public to hand in their overseas greetings telegrams not later than today."

By 1963, the quality and reliability of cable networks had improved, and the birth of satellite communications would see inroads into the traffic levels of the Point-to-Point stations. A Post Office publication entitled 'Radio Communication Services' gives an excellent snapshot of the stations of that particular time, full details being included as details are relevant to the Somerton receiving station, and will assist in understanding the role of the station.

Examples of early 1950s publicity for the service, designed in typically period Post Office style.

The impressive bank of receivers at Baldock, 1952.

"When a telephone or telegraph call is extended overseas, the long-distance Post Office Radio Service is employed if cable circuits are not available. Point-to-Point working is used and the transmitter signal is 'beamed' to a particular receiving station in the distant country. As the transmission is in one directly only, a return path must be provided by a transmitter located in the same country as the receiving station, and the return signal is picked up at a Post Office receiving station.

The transmitting and receiving stations in each country are situated some distance apart and different frequencies are used for transmission and reception. These precautions minimise the possibility of interference between the transmitted and received signals.

Point-to-Point working and a wide range of other services are provided by a number of Post Office transmitting and receiving stations situated throughout the British Isles, see table below:

TRANSMITTING STATION	SERVICES
Leafield	Multi-destination press broadcasts, Point-to-Point Telegraphy.
Criggion	Multi-destination press broadcasts, Point-to-Point Telephony & Telegraphy.
Rugby	Point-to-Point Telephony & Telegraphy, Standard-frequency transmissions, Time Signals, Multi-destination press broadcasts.
Dorchester	Point-to-Point Telegraphy & Facsimile.
Ongar	Point-to-Point Telegraphy & Facsimile.
Bodmin	Point-to-Point Telegraphy & Facsimile.
Portishead	Shore-to-Ship Telegraphy
RECEIVING STATION	
Somerton	Point-to-Point Telegraphy & Facsimile.
Bearley	Point-to-Point Telegraphy, Telephony & Facsimile.
Baldock	Point-to-Point Telegraphy & Telephony, Ship-to-Shore Telephony, Frequency Measuring Station.
Brentwood	Point-to-Point Telegraphy & Facsimile.
Cooling	Point-to-Point Telephony including New York Telephony Service.
Burnham	Ship-to-Shore Telegraphy
Banbury	Frequency Measuring Station
EXPERIMENTAL STATION	
Goonhilly Down	Experimental work on communication via artificial satellites.

These stations provide radio circuits which operate regularly between Great Britain and such countries as Australia, New Zealand, North and South America, the Continent and the Middle East, to mention but a few.

The regular services include such facilities as the transmission of still pictures (facsimile), press broadcasts and standard-frequency transmissions.

In addition to their conventional radio communication commitments, the Post Office is now engaged in research into the problems of setting up a world-wide communication system via artificial earth satellites. An experimental microwave transmitting and receiving station has been established at Goonhilly Down in Cornwall for this purpose.

Two centres in London have technical control of all overseas telephone and telegraph calls. The radiotelephone calls are controlled at Brent Building in North West London, and radio telegraph calls at Electra House on the Thames Embankment.

The Point-to-Point Radio Telephony service provides a link from a home telephone subscriber to someone located overseas. The call from the subscriber is routed from the local exchange to the International Exchange in Wood Street, London, where the call is controlled and timed. Details of the call are taken and then passed to Brent Building where supervision of the call by technical staff takes place. The call is connected by land line to the correct transmitting station and its associated receiving station. The 'go' and 'return' circuits are combined at the International Exchange and are connected to the national telephone network.

Equipment is situated at Brent Building which gives privacy over the radio section of the route. This apparatus splits the speech band from the subscriber into five sections and transposes them so that the messages are unintelligible to any unauthorised person receiving them. The transposed sections of the frequency band are then used to modulate a transmitter. Equipment is installed at the distant receiving station which restores the signal to normal before transmission over land lines to the called subscriber, and similar equipment is installed on the return path to ensure privacy in the return direction.

Other apparatus compensates for differences in speech level in the system and ensures that, within limits, a reasonable depth of modulation is maintained throughout the conversation.

The majority of transmitters operate in the high frequency range between 3 and 30 MHz and employ single-sideband or independent-sideband transmissions. In the independent-sideband system, each sideband carries different information and the carrier wave is transmitted at a relatively low level. The bandwidth of each sideband is approximately 6 kHz and can carry two conversations each of which have a bandwidth of 3 kHz. This means that four conversations or 'channels' can be transmitted simultaneously from each transmitter with an overall bandwidth of 12 kHz.

Directional aerials are used which beam the transmission in the required direction.

Rhombic aerials are used at most transmitting stations because these aerials combine highly directional properties with the ability to work efficiently over a fairly wide range of frequencies. The latter property is desirable as changes in propagation conditions make it necessary to use several different frequencies on one transmission path during different parts of the day.

Point-to-Point telegraphy services are available to all parts of the world and as with the telephone service, the national telegraph network can be extended and messages transmitted by radio. The telegraphy signals from the renters are routed through Electra House where they are converted into a suitable form for radio transmission. Several different types of signalling systems are used to operate distant teleprinters or other devices, but wherever possible, automatic error-correcting signals are being brought into use so that errors due to fading or distortion over the radio path are automatically corrected as the message proceeds.

The Press Broadcast Services is provided for Press Agencies who wish to broadcast news to their agents abroad. The Post Office have transmitters for this purpose operating in the high-frequency range, each of which uses an aerial having a broad beam. The transmissions are thus 'beamed' to large areas such as North America, South America, Europe, and others. This service gives almost world-wide coverage.

Difficulties arise on this type of long-distances transmission due to changes in propagation conditions over the various paths. Forecasts are produced by the Post Office showing which frequencies will give the best service at any particular time for several months ahead. These are sent to the transmitting and receiving stations and the necessary frequency changes are made by both at the appropriate time.

Morse code and Hellschreiber transmissions are at present used for the majority of the Press broadcasts but teleprinter operation is now becoming more popular.

The transmission and reception of still pictures or documents is known as facsimile telegraphy and as with telephony and telegraphy transmission this may take place over normal telephone cables or over a radio path between two countries. The parties interested in this facility are press agencies who require a picture taken in, say, New York for publication in a newspaper in this country a few hours later, or business firms who require accurate copies of documents from an overseas branch.

The picture to be transmitted is scanned by a point of light and the light reflected from the picture on to a photo-electric cell. The electrical output from the photo-electric cell varies according to the light and dark parts of the picture.

The output is amplified and is used to amplitude or frequency-modulate a low-frequency carrier wave. This may be sent over a line circuit to a transmitter where it is used to modulate a high-frequency carrier wave. At the receiving end the reverse process takes place and the signal is used to vary the intensity of a lamp focussed upon sensitised paper. The paper is scanned by the lamp in synchronism with the light spot scanning the original picture. The paper is then developed and after development a copy of the original is produced."

This was, however, probably the last technical advance affecting the wireless service. More efficient and cost-effective methods of transmission continued to develop, using improved cables, satellite services and high-quality landlines.

As cable routes and (especially) satellite communications became more reliable and economical, the future of the wireless stations became uncertain. Many of the stations of course closed due to the pre-war rationalisation programme from 1938, but the remainder maintained in service (although not necessarily a Point-to-Point service) until 2000. Further developments and improvements in communication made significant inroads into the stations, and the increased use of under-sea cables and satellite systems signalled the end of the service towards the end of the 20th century.

Details of each station's commercial opening dates (various sources quote different opening dates depending on the service and network) and associated closures are shown in the following table:

TRANSMITTING STATIONS

Location	Opened	Closed	Notes
Bodmin	20th October 1926	31st March 2002	Initially closed at the same time as the Bridgwater receiving station but became mainly operational under MOD (Navy) contract to the Post Office (ETE/IMTR)/BTI until 2002 (finally under BT Overseas Services)
Leafield	24th April 1922	June 1986	Originally planned to commence operation c.1914 but work was suspended during WW1. However, station was used for experimental and reception purposes during 1914-18. Ceased Point-to-Point services c.1972. Latterly used for PO/BT Maritime Services and training school.
Dorchester	16th December 1927 *(see note)*	July 1979	Services from Dorchester commenced to Rio on 9th August 1927 and New York on 28th October 1927. 16th December 1927 was probably the official opening date. Point-to-Point services ceased in 1970. HF Maritime Radio Services ceased in July 1979.

Ongar	31st October 1921	1985	Ceased Point-to-Point services c.1981. Operated PO/BT Maritime Services after 1975.
Rugby	1st January 1926 *(GBR 16 kHz service)* 7th January 1927 *(Transatlantic R/T service)*	2 August 2007	Last Pt-to-Pt service closed c.1990. Maritime Services ceased 30/4/2000. MSF time signals ceased on 31/2/2007. Temporary LORAN C service operated from 2005 to July 2007. The final 820ft mast was felled on 2/8/2007.
Tetney	8th April 1927	1940	Closed at the same time as Skegness, following the C&W "Concentration" of W/T services. Some equipment relocated to Dorchester.

RECEIVING STATIONS			
Location	**Opened**	**Closed**	**Notes**
Bridgwater	20th October 1926	1940	Closed following the C&W 'Concentration' of W/T services.
Somerton	16th December 1927	30th April 2000	Last Point-to-Point service closed c.1990. Operational until 2000 for BT's HF Maritime Radio Services.
Bearley	1953	1981	Point-to-Point services closed in 1981. Also used for research and development for Post Office Radio Stations.

Baldock	1st June 1929	Still operational	Point-to-Point services closed in 1971. FCS moved into RX building after 1972. Operated by the Radiocommunications Agency then OFCOM. Now an OFCOM radio spectrum management centre
Skegness	8th April 1927	1940	Closed at the same time as Tetney, following the C&W 'Concentration' of W/T services. Some equipment relocated to Somerton.
Brentwood	1921	28th September 1967	Receiving station for Point-to-Point and some Maritime R/T services up to closure. Used as a depot for PO/BT telephones up to 1990.

Today, the point-to-point network is sadly a thing of the past; satellites and cables have put paid to the era of radio communication for commercial use. However, the system served its purpose well over many years, and remains a vital part of international communications history.

CHAPTER 3

THE CHEDZOY EXPERIMENTAL RADIOTELEPHONE STATION

The small, sleepy village of Chedzoy, not far from the North Petherton site of the Beam Station, has a claim to fame which very few people, including local residents, are today aware of. Although not a Beam Station in the true sense of the word, its work proved vital for the development of the long-distance Beam system and the understanding of radio propagation.

The General Post Office had set up an experimental radio station at Three Oaks Farm, Fowlers Plot, just outside of the village, to investigate the possibility of wireless radiotelephone communication over long distances, the selection of the site being described in a local newspaper report:

"In July, 1923, the British General Post Office, in association with the American Telephone and Telegraph Company, the Western Electric and the Radio Corporation of America opened the first Trans-Atlantic wireless telephone office in Britain. In order that the greatest secrecy should be maintained, the sets of instruments, worth £2,000, were installed in Mr. Fry's garage in Chedzoy.

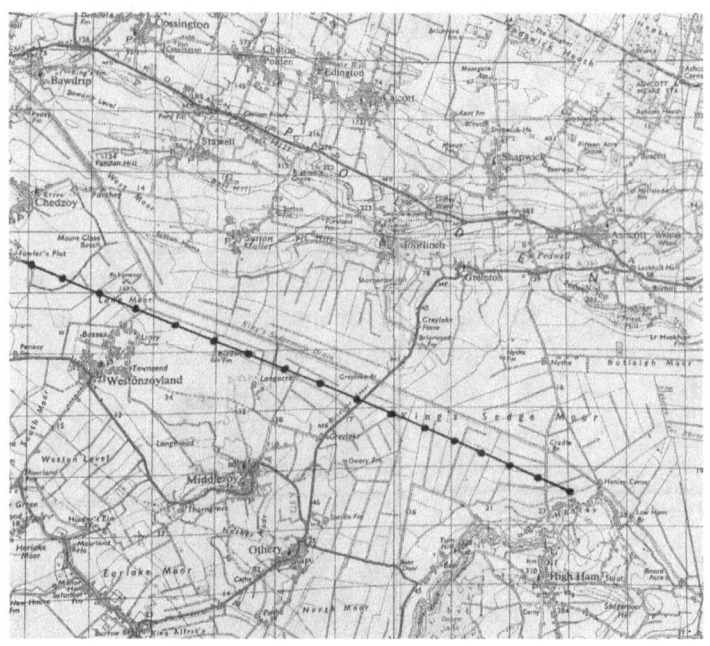

The location of the Chedzoy Wave Antenna,
stretching from Chedzoy to High Ham,
superimposed over a more modern map.

Ever since the Armistice, men classed among the world's greatest wireless experts had scientifically tested every area in the United Kingdom to locate absolutely the finest site possible for the development of international wireless telephonic communication. Thirty months of research ultimately eliminated all claims but those of Chedzoy. This hamlet is below sea level; the surrounding country is flat, and there are no trees in the neighbourhood. The fact that Mr. Fry's farm stands a few hundred yards from the roadway and is circled by rich cultivated land, across which strangers had no reason to pass, ensured that absolute secrecy which the Post Office and companies associating in these experiments sought. But their further reason was that there is some undefined quality in this air peculiarly favourable to wireless transmission and reception.

Whatever this undiscovered quality is, it has the defects of its advantage, as while it intensifies the strength of wireless signals, it has an insatiable appetite for all exposed wires. Even bronze wires about the district are corroded after a short period of service. The Post Office has a special staff of men eternally testing and replacing strands eaten away by this corrosive process.

But because it possessed the manifest advantages mentioned, Mr. Fry's garage at Chedzoy in July 1923, became England's first wireless telephonic station. For the first six months no material success came to embolden the hope that Anglo-American telephonic conversations, such as that of February 8th 1925, would become an everyday affair. Work continued more and more at night.

Then one evening last spring, a British Post Office engineer walked into Mr. Fry's farmhouse and shyly asked Miss Fry if she cared to talk to someone across the ocean. She raced to the garage, and she was the first woman in England to talk to America on the telephone. That conversation last spring marked practically the inauguration of international wireless telephony."

As would be expected, such news caused much excitement in the local press, as a typical report of the time states:

"One of the outstanding new items last week was the fact that wireless telephony with America has been successfully accomplished – accomplished, too, by means of a Wireless Station established almost at our doors; no further off than at Chedzoy, within easy reach, as everyone knows, of Bridgwater. Profound secrecy has surrounded the experiments, which have been going on for some time, the operations being carried out by the British General Post Office authorities in conjunction with the Western Electric Company of America.

In spite of the secrecy, gossip has been busy, with the knowledge that experiments of a wireless nature have been going on; now the reserve has been penetrated, and it is known that telephonic communication has passed through the Chedzoy station with mid-America.

A representative of 'The Mercury' who visited Chedzoy, found two wireless stations, both situated adjoining the road on Mr. Edwin John Fry's farm. One belongs to the British General Post Office, having been converted from a garage; the other – on the opposite side of the road – belongs to the Western Electric Company, America, and consists of a wooden hut, protected by wooden fencing, the interior containing elaborate apparatus used in connection with the experiments. A double length of wire, suspended on ordinary telephone poles, is connected with the stations, and stretches a distance of about seven miles, over Sedgemoor to Henley, near High Ham.

Great reserve is still being maintained by the officials and all concerned, but our representative learned that a married daughter of Mr. Fry – Mrs. Wilkins, of Bridgwater

– who formerly resided with her father at Chedzoy, has actually spoken to someone in America by wireless. One evening, she was asked by a Post Office engineer at the station if she would like to hear a message from across the Atlantic. Instantly accepting, she hastened to the garage, and she received a message from 'the other side'."

Investigation has revealed that 'Miss Fry' would have been Hilda Fry, who married Harold F. Wilkins, in Bridgwater in 1924. It is therefore Hilda who had unwittingly become one of Trans-Atlantic Radio's early pioneers.

Wyndham Fry, one of the sons of Edwin John Fry, recalled shortly before his death in 1998 that the experiments took place over two or three years, and he remembers hearing "Cardiff calling" from the eight-valve radio which was powered by a generator. His sister maintains that she didn't actually hear someone speaking from the United States, but only some noises.

A paper written by E.H. Ullrich, and published in the July 1926 edition of 'Electrical Communication' gives an idea of the equipment used at the Chedzoy station during early field strength experiments:

*Field Strength monitoring equipment similar to
that used at the Chedzoy station, c. 1924.*

The above photograph shows the type of apparatus
that had been used since the summer of 1924 at Chedzoy,
Bridgwater, England, for the measurement of American
stations between the wavelengths of 5m and 18m. The
paper describes the equipment in detail:

"The short wave receiving set is designed to measure
weak fields over the range of 120 to 40 metres. This set,
which was constructed in the Bell Telephone Laboratories,
affords striking testimony of what can be accomplished
by effective shielding. The inner brass box, in which the
apparatus is mounted, has a removable back, held in position
by a large number of screws placed at intervals of about
half an inch. The whole is placed inside a second brass box.
The meter recording the DC output of the thermocouple is
mounted inside the set, so as to be visible from the front.

The current divider, although of resistance wire with minimum inductance and distributed capacity, has had to be calibrated for frequency. The receiving set is a superheterodyne employing four stages of high frequency amplification after the first detector. The 'pick-up' is audible when the full amplification is used, but it is, of course, a constant error, not a percentage, and is quite unimportant even at 40 metres in comparison with the signals that traverse the Atlantic."

Further tests continued throughout 1924 and 1925 to ascertain the accuracy of reception in both directions across the Atlantic. With regard to the reception at Chedzoy, it was reported that:

"Perhaps the most convincing measure of the efficiency of directional receiving systems for transatlantic transmission is the improvement effected in the reception of intelligible words. Results of tests show the improvement which the wave antenna in England has made in the ability to receive certain test words spoken from Rocky Point. For this purpose there was transmitted from Rocky Point a list of disconnected words. A record was made at Chedzoy of the percentage of the words understood for reception on the loop and on the wave antenna. This constitutes a convenient method of rough telephone testing.

It will be appreciated, however, that it would be possible to understand a greater proportion of a conversation than is represented by these results. The curves show that it was possible to receive, for example, 803 of the words for but 9 of the 24 hours on the loop, whereas with the wave antenna reception continued for 18 hours."

The station was heavily involved in numerous tests between 1924 and 1926, far too detailed and technically complex to recount in any great depth; suffice it to say these comprehensive tests proved vital in understanding the nature of radio propagation on short wavelengths.

The Rocky Point (USA) aerial system which communicated with Chedzoy.

An article in the May 1926 issue of *'Experimental Wireless'* however, gives a concise overview of the tests which took place at Chedzoy and other stations on both sides of the Atlantic Ocean:

"A report upon two years' measurement of Transatlantic signals, etc., preparatory to the arrangements for commercial telephony.

The measurements were initiated to show the conditions obtaining throughout the twenty-four hours of the day and the various seasons of the year. The work was done under the auspices of the American Telephone and Telegraph Company, The Bell Telephone Laboratories, and the Radio Corporation of America on the United States side, and on the British side by the International Western Electric Company and the G.P.O.

In America, measurements were made at three stations on signals from Northolt on 52,000 cycles, and from Leafield on 34,130 cycles and 24,050 cycles. In England, measurements were made at New Southgate and Chedzoy, Somerset, on the transmissions from Rocky Point 2XS on 57,000 cycles, Marion WSO on 25,700 cycles and Rocky Point WQL on 17,130 cycles."

The Postmaster General was prompted to issue a statement following incorrect press reports regarding the Trans-Atlantic radiotelephony tests:

"Statements have recently appeared in the press on the subject of Trans-Atlantic wireless telephony to the effect that two-way telephonic communications have already been established between England and America by means of a new and secret system. The Postmaster General desires it to be understood that this is not the case. The experiments so far are a continuation of those announced to the press in May 1924."

This statement would therefore seem to indicate that the reports of successful two-way communication with America were exaggerated by the press; however, there is no doubting the usefulness of the experiments conducted by the station.

It was subsequently reported that the equipment used for the experiments cost in the region of £2,000.

Another report couldn't contain the author's wonderment over the 'marvellous' new invention:

"England and the United States have spoken to each other on the marvellous new wireless telephone. The best wireless telephone station in England is on a farm owned by a Mr. Edward Fry, at Chedzoy, near Bridgwater. There are three stations, one at Swindon and two at Chedzoy, both on the farm of Mr. Edward Fry."

The station also took part in experiments with regard to directional receiving aerials, something which would be used to great effect by the nearby North Petherton Beam Station. One such experiment took place during 1923 and 1924 with regard to the reception of signals from Cairo (call sign SUC) using the Chedzoy Wave Antenna.

Another report was published in September 1924, which makes for interesting reading:

General

Advantage was taken of the Wave Antenna at Chedzoy to carry out tests on the reception of Cairo and to compare with the commercial reception at Banbury.

The Wave Antenna as normally used for Trans-Atlantic reception consists of a double line but for the Cairo reception, where the signals are incoming from an opposite direction, a single line was used, both ends being grounded through suitable resistances.

The usual long wave receiver was employed with, at times, further low frequency amplification. A tuned low frequency amplifier was also tried during periods of heavy atmospheric disturbance but its usage was not continued, mainly owing to the variability of the frequency of SUC, although it was found that the ringing set up by the atmospherics in this receiver made aural reception very difficult.

Tests

The tests were carried out in two parts. In the first, which extended from 25/9/23 until 12/10/23, the signals were directly received at Chedzoy and the results compared with those received at Banbury. This period was one during which atmospherics and fading were not troublesome and the results given in Table 1 below show little difference between the reception at the two stations, signals at Chedzoy being very much stronger than those at Banbury, however, they were much more easily read.

TABLE 1	Messages Sent	Received Complete	RQ's	RQ's %
Chedzoy	645	626	19	3.0
Banbury	645	620	25	3.9

Note: RQ's – Requests for Repetition.

In the second part of the tests and which extended from January until August 1924, the signals were relayed from Chedzoy over the land wires to the C.R.O. (Central Radio Office, London) where they were aurally received. During this period, reception was for the main part on Monday evenings between the hours of 18.30 - 21.30 GMT. This time is a period over which Cairo's signals are weak and atmospherics relatively strong.

A detailed analysis of the results was rendered difficult by land line troubles which were frequent. Table 2 below gives a summarised analysis based on the number of messages requiring RQ's.

TABLE 2	Messages Sent	Received Complete	RQ's	RQ's %
Chedzoy-C.R.O.	1019	905	112	11.2
Banbury	1282	1141	141	11.0

Conclusions

Since the direction of maximum intensity for atmospherics varies over a sector which includes the direction London-

Cairo it was not possible to expect the same effectiveness with the Wave Antenna as could be obtained with signals from America.

The results indicate that whilst remote reception of Cairo is possible using the Wave Antenna, no effective advantage in the reception as compared to the direct reception at Banbury was likely.

In general it was found that the advantage in Signal Stray Ratio which accrued at Chedzoy from using the Wave Antenna was balanced by noises on the land line circuit to the C.R.O.

In January 1925, the station was involved in pioneering tests of radio propagation during a solar eclipse, details of which were reported as below:

"During the solar eclipse of Saturday, January 24, the engineers of the Post Office and of the International Western Electric Company, at the instance of Admiral Sir Henry Jackson, measured the strength of wireless signals received from New York at the Chedzoy station in Somerset and at New Southgate, London. The results show that the signal strength rose to a sharp maximum and fell to a very sharp minimum during the progress of the eclipse.

At 14.12 GMT, totality occurred at New York and more than half of the Trans-Atlantic track of the signals was in partial darkness; at this instant the strength of the signals received in England was observed to be about twenty per cent above normal. At 14.52 first contact was visible in

London, and by this time signals had increased to about double the normal strength.

A few minutes later the centre of the total phase was in mid-Atlantic about 400 miles to the south of the wave track, and the whole of the track was now in partial darkness. The signal strength rose to a maximum, first at Chedzoy and then at New Southgate.

During the next half-hour the centre approached the wave track rapidly by moving in a north-easterly direction, and signal strength decreased greatly. At 15.30 the last contact occurred at New York, and at about 15.40 the signals received at both places of observation had fallen in strength to a minimum value less than one-fifth of the normal. At this instant about 300 miles of the western end of the wave track was in full daylight and the centre of the eclipse was crossing the wave track about 500 miles to the west of Ireland. By 15.45 the centre had moved beyond the Faroe Islands, daylight had returned, and at about 16.10 normal signal strength was regained.

Throughout the eclipse directional measurements were made by the staff of the Radio Research Board, but no effect on bearings could be detected."

There was still a great deal of wonderment locally about the fact that someone could make a telephone call to the United States by using a small wireless station located in the Somerset countryside. In fact, on April 18th 1925, it was reported locally thus:

Receiving and ancillary equipment at the Riverhead receiving station, which again would have been similar to that installed at Chedzoy.

By 'phone to the States. Wireless from Bridgwater to Mid-America. A successful experiment

"A telephone call was put through from Hampstead to Mid-America. The conversation from London was conveyed by an ordinary trunk line to a point near Bridgwater and wireless to Rocky Point, where it was picked up and conveyed to the American subscriber's home. The reply was carried back in the reverse order, and the conversation was plain. Thus wrote a correspondent of the *'Morning Post'* on Wednesday. The article continues:

When that North London subscriber spoke to mid-America on February 9th, a series of connections ran from the subscriber's telephone to the local exchange, thence to the central switchboard in London, where the line was plugged on to a special trunk cable from London, through Bristol, to Bridgwater, thence by another special landline across the three miles to Chedzoy, where it was wirelessed to Rocky Point, and thence by landlines to the American subscriber's mid-Western home. The reply came back by the same way.

The call was not made through England's first experimental wireless telephonic station near Swindon - it was put through by a second station, Chedzoy, near Bridgwater. To be precise, there are three stations, one at Swindon and two at Chedzoy, within the length of a cricket pitch of each other. They are both on the farm of Mr. Edward Fry."

'Wireless World' reported on October 8th 1925 that:

"The imaginative speculations of the novelist and the newspaper Press as regards the possibility of a public telephone service between this country and America are nearing realisation. Transmission tests are likely to take place between the Rugby station and America within the next few days. The International Western Electric Company, who have been responsible for the plant at Rugby, have now handed over the installation to the Post Office. The receiving station is situated at Chedzoy, in Somerset, and for several months messages have been successfully received from Long Island, N.Y. According to the Morning Post, the charge for a three-minute call between London and New York will probably not exceed £1."

Under the headline of 'Talking to America', another press report of the time proclaimed much excitement about the new wireless telephony link to America:

"There are now available facilities to enable a person in England to talk by wireless telephone to an American in New York or elsewhere in the streets, and vice versa, and only awaiting is the official exclamation, "Go!"

Paying, say, £1 for a three-minute 'call' to New York, the 'caller' will be switched to the new gigantic telephone transmitting station at Rugby, and thence his message will be despatched across the Atlantic as easily as though he would speak to a friend or merchant a street or two distant. On the other hand, the New York to England 'caller' will have his message across the Atlantic received at Chedzoy, Somerset, and thence it will be telephoned by the usual land system to London or other city or town. Another marvel of science is such a link between country and country separated by a vast stretch of sea."

Details of the work carried out at Chedzoy were promulgated later that month:

"In discussion with a Press representative, an expert stated; nobody prophesies too much where wireless is concerned, but it certainly looks as if commercial operation will be feasible very much more quickly than many appear to think. There is little difficulty now in transmitting or receiving a wireless telephone call across the Atlantic, but have still to get away from freakish results and achieve something that is dependable enough at all moments for commercial requirements. What may be quite good enough for enthusiastic amateur is not good enough for a service that has a commercial basis. We are undoubtedly near the time for important development, and the Post Office is applying itself busily to the problem.

Reception of American test messages has taken place at a Post Office Station, also constructed by the International Western Electric Co., at Chedzoy, Somerset. When the London subscriber telephones to New York his message will be transmitted via Rugby, and when American subscriber telephones to London the call will pass through Chedzoy. The two stations will work on different wavelengths."

Numerous tests between the Chedzoy station and its counterparts in the USA continued to be conducted throughout the year, and various papers were written and published in the technical journals of the time.

Some of these papers were extremely complex, but nonetheless contributed to the understanding of radio propagation and the effects of day and night combined with the choice of wavelength in use; something which would be used to great effect by both point-to-point and broadcasting stations over the next few decades.

Various formulae and graphs were produced to explain the link between received field strength of radio signals with wavelength in use at the time, which were used to select the optimum time and wavelength when the commercial service between the USA and the United Kingdom became operational. The findings were also used to assist in wavelength selection between the United Kingdom and other countries involved in the Beam Wireless service.

It therefore fair to say that the Chedzoy station contributed in no small way to the development of long-range radio communication; and as we shall see, tests continued to take place over the next couple of years.

It is interesting to discover the views on the new service from the American side; one magazine seemed somewhat scathing as regards to the tariffs charged for the service:

"The large station at Rocky Point (Radio Central) has for some time been able to send telephone messages across the Atlantic to England where a receiving station has been set up at Chedzoy; this, it is expected, will be the English receiving station for the Trans-Atlantic radio-phone channel. The British have been at work on a transmitting station for their end of the channel and a cable to the New York Times says that the station is now complete and has been taken over by the British Post Office, which will operate the communication scheme.

The English news puts the price of a three-minute talk to America at five dollars. This seems like an unreasonably low price for the service and certainly cannot be based on the idea of earning a reasonable return on the investment."

The experiments were completed in early 1927, and the official report issued by the Committee on Transatlantic Wireless Telephony on 27ᵗʰ May 1927 is significant:

Terms of Reference

We were appointed by the Postmaster-General in March, 1923, with the following terms of reference: "To consider in the light of recent progress in wireless science the possibility from a technical standpoint of transatlantic wireless telephony of sufficient reliability for commercial use, and to advise what practical steps, if any, can at present be taken to develop this means of communication."

Progress in Wireless Science

The development of the thermionic valve and its application to wireless transmission and reception had opened up possibilities of transoceanic wireless telephony which had been beyond practical attainment previously. In 1915 the American Telephone and Telegraph Co., by connecting together a large number of small valves which were then available, succeeded in transmitting speech across the Atlantic from Arlington, U.S.A., to Paris, and between 1915 and 1923 developed what is known as the "single-side-band suppressed-carrier" method of working in wireless telephony, an invention which is designed to reduce the number of wavelengths required for telephonic communication and at the same time offers very considerable advantages in the utilization of power required for transmission.

Attention was also paid to the development of a large-power valve, and to the production of a water-cooled valve of 10 kilowatts continuous and reliable output. These advances, together with a large number of other important improvements, opened up possibilities of long-distance wireless telephony.

With larger valves and improved technique, the company again achieved one-way transatlantic telephony, and on the 15th January, 1923, a large audience at the works of the Western Electric Co. (now Standard Telephones and Cables, Ltd.) at New Southgate, London, heard the voices of the speakers in New York quite clearly and loudly, speaking at predetermined times over a period of two hours.

Preliminary Experimental Work

The preliminary work of the committee was devoted to obtaining data for outlining the problem upon which it was engaged. It was known that wireless signals across the Atlantic varied in strength from hour to hour and day to day, and that the disturbance due to what are known as atmospherics varied also at different times of the day and at different seasons of the year.

The American Telephone and Telegraph Co., the Radio Corporation of America, the International Western Electric Co. (now International Standard Electric Corporation), and the Post Office Engineering Department offered their co-operation in the necessary experiments.

To obtain the required data it was decided to make measurements in this country of the strength of signals from America, and the strength of the atmospherics, each weekend, over a complete twenty-four hours in each case. These measurements were continued to the end of 1926, and numerous records of the conditions likely to be met in giving a commercial service were thus obtained. At the same time the Post Office wireless stations, Leafield, Northolt, and latterly Rugby, transmitted signals to America, where similar measurements on signal strength and atmospherics were also being made.

The special signals from America were sent from the Radio Corporation station at Rocky Point, Long Island. Speech was also transmitted from this station, and measurements of the deteriorating effect of atmospherics on the intelligibility of speech were made in this country.

Arrangements were also made concurrently for the Post Office Engineering Department to install at Chedzoy, Somerset, a receiving antenna, seven miles in length, for the purpose of gaining experience of the value of this type of antenna in combating atmospheric disturbances.

Installation of a 200 kW Telephony Transmitter at Rugby Wireless Station

After the preliminary data from these measurements became available it was possible to decide upon the size and type of transmitter and the receiving arrangements which would be necessary in order to attempt a commercial telephone service between England and America, and the committee recommended the Postmaster-General to install a 200-kilowatt telephony transmitter at the Rugby Wireless

Station for this purpose, and in order to see what further difficulties would be encountered. It was hoped that this transmitter would, if successful in the experimental stage, be suitable for commercial working later.

The telephony transmitter apparatus was ordered by the Post Office from the Western Electric Co. (now Standard Telephones and Cables, Ltd.), who installed it at Rugby. A separate antenna and earth system and the necessary oscillating circuits for the telephony installation were also erected by the Post Office.

It was thus possible to obtain the advantages of conducting the initial experiments at the Rugby Station, where the high masts and the necessary power and certain plant were already available without additional expense.

The telephony transmitter installed at Rugby comprised modulation equipment and power amplifiers. The modulation equipment converts the speech currents arriving over the cable system from London into weak wireless signals, using the "single-side-band suppressed-carrier system" already mentioned.

After the wireless signals are produced in the modulation equipment they are amplified in three stages for transmission from the antenna. The power-amplifier consists of thirty water-cooled metal-glass valves, each of 10 kilowatts rating.

This transmitter was designed on somewhat similar lines to that which was already in use by the American Telephone and Telegraph Co. at the Rocky Point Station, full advantage being taken of the experience gained in the working of that installation. The British installation was ready for trial early in 1926, and, after preliminary tests had been made, two-way conversation between England and America was first attained on the 7th February, 1926.

Installation of Receiving Arrangements at Wroughton and in the United States

Improvements had also been made in the receiving arrangements. In this country a double receiving-antenna had been erected at Wroughton, near Swindon, in order to be near the main underground telephone system from London to the West of England.

Measurements and tests carried out in Scotland in 1925 and 1926 indicated that a considerable improvement would be obtained by reception in the north, and another receiving antenna is being built at Cupar, which it is expected will be ready shortly.

In America also experiments had been made, using a receiving antenna at different locations. As a result it was found that the more northerly the location the greater was the freedom from atmospherics, and the American Telephone and Telegraph Co. built a receiving-antenna system at Houlton, Maine, near the Canadian border. Improvements on this antenna, and the receiving arrangements have been continued, and further developments which have been suggested will be put into effect.

Decision to operate both Transmitters on the same Wavelength

The preliminary two-way conversations were made on two separate wavelengths, and had to be confined to week-ends, so as not to interfere with other wireless services. The early measurement work on signals and atmospherics had indicated that the optimum wavelength for this transatlantic communication would be between 5,000 and 6,000 metres. In this band there were over forty wireless telegraph services already working, and the difficulties of finding room in this crowded range of communication were such that it was decided to attempt to operate both American and British transmitters on the same wavelength.

This introduced a difficult problem, which was solved independently on both sides of the Atlantic by the development of switching devices actuated by the voice. These devices switch the transmitting and receiving circuits at each end on or off, according to the direction in which the speech is proceeding.

Clearing the Waveband of Existing Wireless-telegraph Services

Having obtained satisfactory two-way telephony on one wavelength, it was decided to ask the authorities responsible for certain wireless telegraph services whether they would be willing to alter their wavelength, and we have pleasure in recording the friendly spirit of co-operation shown by the Government Administrations in Germany, Italy, and Russia, and by the Air Ministry and Marconi Co. in this country, in making these changes in order to make room for the transatlantic wireless telephone service. On the American side also there were a number of wireless stations in the waveband desired, under the control of the United States Navy, and they also kindly altered the distribution of wavelengths used by their stations.

Opening of a Commercial Service

In October, 1926, the experiments had progressed so far as to show that reasonable commercial working with America should be possible over a considerable portion of twenty-four hours each day, and the committee recommended the Postmaster-General to establish a commercial service on an experimental basis, in order to ascertain those defects and possible improvements which can only be eliminated and secured respectively by operating under actual service conditions. After consultation between the Post Office and representatives of the American Telephone and Telegraph Co., the service was opened for public use on the 7th January, 1927.

The operating hours were limited to the afternoon in this country, the portion of the day in which the business hours overlap those of New York; but with the expansion of the service it would be possible to extend considerably the hours of operation.

Further Work recommended

The telephony service between England and America is not secret in the full sense of that word. It was considered, however, that in spite of this a public telephone service between England and America would be of such value that it should not be withheld until a secret system became available. We recommend, however, that the improvement of the present partially secret system of wireless telephony should be pressed to a conclusion by the Post Office engineering staff in consultation and co-operation with the American Telephone and Telegraph Co.

The work of the committee has now been completed, and it is pleased to be able to report that, as a result of the necessary research and experiment, it has been found possible to overcome the initial technical difficulties and to establish a transatlantic wireless-telephone service of sufficient reliability for commercial use. Further technical and operational development will be necessary, and can be left to the Engineering and Traffic Sections of the Administrations on both sides.

The committee desires to place on record its cordial thanks for the co-operation given by the American Telephone and Telegraph Co. and its associated company, the Bell Telephone Laboratories, the Radio Corporation of America, the International Standard Electric Corporation, and the Post Office Engineering Department. The committee also desires to express its appreciation of the assistance rendered by Lieut.-Colonel A. G. Lee, who has acted as secretary to the committee, and under whose direction the necessary data have been obtained, and who also visited the United States to complete the co-ordination of the necessary engineering arrangements by which success has been achieved."

Upon the completion of the Trans-Atlantic testing, the Chedzoy site was initially deemed surplus to requirements.

However, the Post Office had other ideas for the site, as reported locally later in 1925, although the text does indicate an unsurprising lack of understanding of the system:

"By the autumn of last year the experiments at Chedzoy demonstrated to the satisfaction of the Post Office and the associating companies that an Anglo-American Telephonic Service was possible. But would similarly satisfactory results be achieved in a location where the atmospheric conditions were not as favourable as Chedzoy?

Experience alone would give the answer. Consequently it was arranged for the experimental station to be transferred to the neighbourhood of Swindon, where the conditions are almost precisely the reverse to those of Chedzoy. But in order that the continuity of the work in Somerset should not interrupted, it was agreed that the Western Electric Company should open a second station on Mr. Fry's farm.

Early last winter a tiny hut was built adjacent to the Post Office station, and just before Christmas a supplementary set of instruments was set up in this little building, which to all appearances was simply an ordinary farm outhouse. Then, ten weeks ago, a motor-lorry drove up to the garage. The Post Office experimental set was loaded into it and conveyed to Swindon, so for the moment only the Western Electric Company station was operating there.

I managed to find it this (Tuesday) morning, writes the *'Morning Post'* special correspondent from Bridgwater, though only after hard search and a long plough through sodden fields. I had hoped, earlier, that it would be easy to identify it. But the aerial is a faithful replica of the ordinary overhead telephone wire. It has this distinction: It is seven miles long. It stretches from this rural seclusion across the battlefield of Sedgemoor, on past Weston Zoyland, and across the moors and farms to Henley.

The network of wires, running from 30 ft. masts to the respective huts, is cunningly concealed, and visible only when you are close to the huts. I was able inspect the interior of the Western Electric Company's hut - but the Post Office station was locked and shuttered convincingly. In practically all things the apparatus of the Western Company's hut here is faithful model of that in the Post Office Station near Swindon.

Latterly a series of thirty-six hour tests have been made during the weekends. I understand that the results have been so successful that it is probable that the Swindon station will be suspended or abandoned in the near future. Corroborative evidence of this lies in the fact that not only is the Post Office retaining Mr. Fry's garage, but - 1 am credibly informed – the Post Office has decided to keep the garage ready for re-occupation now at the shortest notice.

For transmitting the human voice from England across the Atlantic, the Western Electric Company here obtains the power from secondary cells, which are recharged periodically in Bridgwater. There is not a single source of primary power about the place.

When that North London subscriber spoke to Mid-America on February 8th, a series of connections ran from the subscriber's telephone to the local exchange, thence to the central switchboard in London, where the line was plugged in to a special trunk cable from London, through Bristol, to Bridgwater, thence by another special landline across the three miles to Chedzoy where it was wirelessed to Rocky Point, and thence by landlines to the American subscriber's Mid-Western home. The reply came back by the same way.

But how do those engineers throw the voice from ordinary telephone lines out into the air and across more than three thousand miles to Rocky Point? A delicate and wonderful instrument called the transformer supplies the magic. The incoming voice is caught on that aerial, which stretches seven miles across land from the Western Company's station. Here it flashes to the station, and the transformer brings it again in to the wire. It starts by ordinary telephone line to Bridgwater and on this special trunk line to London.

This trunk was not specially constructed for these experiments. It was merely the best trunk line available, safeguarded by such perfect organisation that interference with it by any other underground cable other source was practically impossible. It is understood here that the central wireless telephonic exchange connecting Britain with America will be in the neighbourhood of Bridgwater. It is not certain that messages have to go thence through Rugby.

I am informed that with Marconi's perfected development of short wavelengths, there is necessity for use of a high-power station to flash the spoken words to America. Meanwhile it is significant that the Marconi Company have decided to construct a short wavelength beam station at North Petherton, which is almost at the very door of Bridgwater, and have also decided to erect a relay station at Highbridge, which is only six miles away from Bridgwater!

It is hoped, however, that the co-operation of none of these stations will be necessary, and that English accents will float direct from the local wireless telephonic station to the United States. As at Swindon, the most intensive precautions are taken here to preserve secrecy over these experiments. The reason for this is not apparent."

Although the wireless telephony experiments had been successfully completed, it was not quite the end of the use of the Chedzoy 'Wave Antenna' During April, May and June 1928, the aerials at Chedzoy and the Marconi aerials at the maritime receiving installation at Highbridge undertook numerous comparison tests, details of which appeared in a detailed Post Office document. The tests, involving two ships, the *'Rajputana/GLWV'* and *'Cedric/GLSM'* were very comprehensive, as extracts from the official report show:

"The object of these experiments was to compare the Chedzoy Wave Antenna system with the Marconi system employed at Burnham Radio, in the reception of traffic from distant Atlantic and Mediterranean ships.

To obtain a direct comparison of the two systems, tests were arranged with two boats, the S/S *'Cedric'* bound from Liverpool to New York, and the S/S *'Rajputana'* bound from Southampton to Yokohama (via the Suez Canal). The transmissions consisted for four tests per day from each boat, two on 2013 metres and two on 2100 metres. Each test consisted of 5 minutes calling, 1 minute dash and 10 groups of 10 letter code, known only to the ship's operator.

The transmission times were as follows:

	Time (GMT).	Wavelength in metres.
S/S Rajputana/GLWV	1100	2100
	1148	2013
	2200	2100
	2218	2013
S/S Cedric/GLSM	1020	2100
	1048	2013
	2300	2100
	2318	2013

The test messages were received at Burnham and Chedzoy, and the results obtained compared with the original messages forwarded by the ship's operator.

The Chedzoy receiving system

The Wave Antenna employed consists of two copper wires supported on telegraph poles and the length used for these tests was approximately 4½ miles.

The receiving hut was situated at the western end of the line and the direction of the line was roughly east and west, the great circle direction passing through New York, Chedzoy and Constantinople.

The receiver is of the superheterodyne type, and to save time, the intermediate frequency amplifier of a standard seven valve Burndept Broadcast Receiver was employed. It was, however, considerably modified for this work, by the addition of separate heterodyne oscillators and filters. The intermediate frequency of the receiver was 48.92 kilocycles.

The Burnham receiving system

Three receivers of Marconi design are employed to cover the range 1050 to 2600 metres. The receivers employ the Bellini Tosi type of directional reception, each one being associated with a separate aerial system consisting of Bellini Tosi loops mounted on 100 ft. steel tubular masts. The receivers are arranged for 'open' figure of eight or 'cardioid' reception followed by 5 stages of neutrodyned H.F. amplification, one rectifier stage, two stages of L.F. amplification, and an optimal note filter. The local heterodyne oscillator is coupled to the rectifier stage.

Results

From S/S *'Cedric'*, the results obtained at Chedzoy were slightly better than the Burnham results. No readable signals were heard at Burnham after the boat was 4 days out, approximately 1154 miles from Burnham, while percentages of 78 and 75 were obtained at Chedzoy on the fifth day when the boat was approximately 1325 and 1558 miles respectively from Burnham. The average percentages for the first five days were Chedzoy 85.1 and Burnham 77.4.

From S/S *'Rajputana'*, the results indicate that up to June 25th (boat 500 miles W. of Alexandria), there was not much difference in the results from the two systems. On June 25th, however, the morning transmissions were not heard by Burnham while percentages of 95 and 100 were obtained at Chedzoy.

The evening transmissions were heard but unreadable at both stations. On the 26th June the morning transmissions were unheard at both stations, but Chedzoy showed a decided improvement on the evening transmissions. No further transmissions were made after June 26th. The reception percentages for the period ended June 26th were Chedzoy 74.2 and Burnham 65.5.

Conclusions

The results indicate that good readable signals could be obtained at Chedzoy from both ships when the signals were unreadable and even unheard at Burnham, and that by the installation of such apparatus it should be possible to obtain one day additional contact on West-bound ships and about two days extra contact with East-bound ships. In reaching this conclusion account has not been taken of the following points:

- The Chedzoy receiver could be considerably improved if it were re-designed and reconstructed as a complete unit for this service.

- In the foregoing tests, the Chedzoy receiver had to be used on two different wavelengths and tuned and balanced up for each transmission whereas at Burnham separate receivers were available on each wavelength and could be left in adjustment.

- The operator at Chedzoy was an ex-R.A.F. operator with less than two months service in the Post Office and was totally unfamiliar with the service.

- As regards East-bound traffic, the Wave Antenna was directed 10 degrees to 15 degrees north of the actual direction of reception with consequent diminution of received signals."

A record of the Trans-Atlantic wireless telephony experiments over the early years of the service indicates the importance of the Chedzoy station.

Years	Transmitter (USA)	Call sign	Frequency	Receiving Station (UK)
1923-1924	Rocky Point	2XS	57 kc/s	New Southgate
1925	Rocky Point	2XS	57 kc/s	Chedzoy
1926 (to Sept.)	Rocky Point	2XS	57 kc/s	Wroughton
1926 (Oct.)-1928	Rocky Point	WNL	60 kc/s	Wroughton
1927-1933	Rocky Point	WNL	60 kc/s	Cupar

As the wireless telephone service to the United States became more established, new receiving aerials were erected at Wroughton, near Swindon, providing a more reliable land-line connection to London and at Cupar, Scotland, which made the Chedzoy site (which was only initially erected as an experimental station) redundant. It had served its purpose well, and its place in history is assured. However, it soon became apparent that this area of Somerset was extremely well-suited for long-distance radio communication, and with this in mind a new receiving station was constructed at Somerton, only a few miles from the Chedzoy location.

The local press were keen to report this new station, and an article dated 24th April 1927 gave details:

"An important wireless station is now in course of erection at Somerton by Marconi's Wireless Telegraph Company for the reception of wireless messages from New York and Rio de Janeiro, the engineer in charge of the work being Mr. H. M. Burrows.

The site is on Black's Moor Hill near Highbrooks Farm, north of the Langport and Yeovil main road, approximately 165 acres being purchased for the purpose principally from Lord Ilchester and Mr. R. W. Pretor-Pinney. It will be entirely a receiving station and situated about half way between the wireless stations at Dorchester and Chedzoy.

The installation is described as a double receiving station operating on the beam 'direct wireless' principle and the doubt aerial system will consist of two lines each of five latticed steel masts 277 feet high each with a cross arm at the top 90 feet long. The length of each aerial will be about half a mile. Near the junction of the mast lines which will form practically an obtuse angle, the cabin of the station will be erected. It is intended to erect a building which will be of considerable dimensions in two parts connected by a short corridor.

From the cross arms of the masts a system of vertical wires will be connected to the building by copper tube conductors carried on metal posts about two feet above the ground, and this system of copper tubes will extend nearly the full length of each line of masts. The building itself will contain the receiving apparatus for the two lines of aerials together with a small charging plant driven by an internal combustion engine of about 15 horse power.

The building will be connected with the Company's London Telegraph Office by telegraph lines and messages received from North and South America will be transmitted over the telegraph lines to Radio House to be recorded there and passed on to their respective addresses."

The opening of the Somerton and Bridgwater Beam Wireless Stations signalled the end of the Chedzoy experimental site, with its pioneering work completed. However, the small village can be proud of its valuable (if short-lived) contribution to the development of radio communication, even if its role may not have been as well-known as some of its peers.

The local press in Somerset recalled the momentous events of the station in a short '25 years ago' section which featured on 15th April 1950:

"Miss H. Fry, of Chedzoy, near Bridgwater, became, last spring, the first woman England to talk to America on the telephone. She spoke from her father's garage, which became England's first wireless telephonic station."

Nothing now remains of the actual 'Wave Antenna' or the supporting masts; and the buildings have now returned to their original use as farm outbuildings, with no evidence or recognition of the significant part the station played in radio history.

Hilda Wilkins (nee Fry), who it is claimed made the historic transmission described earlier, continued to live in the Bridgwater area until her death on 31st March 2008, at the remarkable age of 106. She was at the time regarded as the oldest living person in Somerset.

A recent view across land at Chedzoy, close to where the 'Wave Aerial' would have been located.

A recent view of Oak Tree Farm, Chedzoy.

CHAPTER 4

THE CONSTRUCTION OF THE BRIDGWATER STATION

One of the first mentions of the new station appeared in the local press in February 1925. Under the heading 'Wireless Project', the article read:

New station proposed near Bridgwater, farm and lands to be acquired

Two matters of much interest came before the Bridgwater Rural District Council at their fortnightly meeting on Wednesday. One was a proposal by the Government to acquire over 200 acres of land at North Petherton for the purpose of establishing a wireless station, and the other was a housing scheme for the rural area.

The proposal to establish a wireless station near Bridgwater came before the Council in a letter received from the General Post Office, London, stating that the Postmaster General was proposing to purchase Copse Farm (203 acres) in the parish of North Petherton, and also the four fields intervening between the Copse Farm property and Huntworth Lane. The property was being purchased for use as a wireless station.

The Ordnance Survey map showed a footpath starting from Huntworth Lane, crossing a field and Copse Farm, and terminating apparently at Petherton Park Farm, and if the property was required for use as a wireless station it would be necessary to stop up and prevent access over this footpath. From information obtained from local residents it would seem that that the footpath was purely a private track for the benefit of Copse and Petherton Park.

The Council were asked where whether they agreed that there was no public right of way over the footpath and there was no objection to its being stopped up. The owner of Petherton Park offered no objection to such stopping up.

Mr. Slocombe said he did not wish to influence the Council in the matter at all but personally he did not think there was a footpath there. All the property used to belong to one landlord, and he believed the path was simply one for the convenience of getting to Copse Farm. It had no outlet. He had seen a map of North Petherton printed in 1840, and there was certainly no footpath shown on the map.

The Chairman said he did not think they could say that no path existed without making some enquiry.

Mr. Young thought a small committee should view. The Clerk said it was a parish matter, and he suggested the Parish Council being asked if they had any reason for objecting to the path being closed. Mr. Berry thought it was rather more than a parish matter.

Mr. West proposed and Mr. Greenhill seconded, that the question be referred to the Parish Council. The Chairman agreed with the suggestion of Mr. Berry that the Parish Council should call a public meeting and give every opportunity for inquiry, and then report to the District Council.

The motion was agreed to."

16th March 1925 saw the first mention of the new station in Parliament:

Sir H. BRITTAIN asked the Postmaster-General whether he is now able to inform the House as to the position of the site selected for the Beam Wireless station for communication with South Africa.

Viscount WOLMER: A site near Bodmin for the sending station for the beam service with Canada has been placed at the disposal of the Marconi Company, and it is anticipated that a site near Bridgwater for the corresponding receiving station will be available in the course of a few days.

The same sites will be used for the stations for communication with South Africa, and an order for these stations will be given as soon as official information is received concerning the erection of corresponding stations in South Africa.

A further article appeared locally later that month. Under the heading of 'North Petherton Wireless Station – Important Government Project', the interesting article stated that:

Communication with Dominions

It will be noted from our report of the Bridgwater Rural District Council meeting elsewhere in this issue that a communication was received from the Postmaster General in reference to the proposed purchase of Copse Farm, North Petherton, comprising 203 acres, with four adjoining fields, for use as a wireless station. It is understood that should the scheme mature, this station would be operated on the wonderful new Beam System, which constitutes one of the most important developments of wireless telegraphy. The Government is providing a number of these stations by arrangement with the Marconi Company for the purpose of establishing special communication between this country and the Dominions, including Canada and Australia, and if the North Petherton station is included in this scheme the apparatus required would necessarily be of a very powerful type with huge aerials, masts, etc.

The local Post Office authorities have no information available concerning the proposed station, and at present it is too early to make any detailed statement, but it may be pointed out that in any case there would be no interference with the local reception of the ordinary programmes of the B.B.C.

The Petherton station would have no connection with that at Burnham, which is one of a series being erected around the coast for communication with vessels at sea, or with the wireless operations at Chedzoy (in the Chedzoy district).

The question has been raised of whether a footpath which crosses the property in question is a public or private one, and this has been referred to the Parish Council, but it is considered unlikely that any objection will be forthcoming on this ground.

It was clear that the Council viewed the possibility of this station as an exciting project, and even the local newspaper referred to the scheme as the 'Great Wireless Project' in an article from May 1925:

Government select site near Bridgwater – one of a chain of Empire Stations

It now appears to be an assured fact that the Government intend erecting a Great Wireless Beam Station at North Petherton, near Bridgwater, some particulars of the project having appeared in our last issue.

Sir William Mitchell Thompson, Postmaster General, announced to Parliament on Tuesday that the sites for the Canadian and South African Wireless Beam Stations had been selected.

They would be near Bodmin and near Bridgwater, and it was hoped that the negotiations for the purchase of the particular properties would be completed in time to place the work in the hands of the contractors by the end of the present month.

The site selected at North Petherton is Copse Farm, comprising 203 acres, with four adjoining fields, and the carrying out of the scheme will be one of much interest to Bridgwater and the surrounding district.

An Empire Scheme

The first site at Bodmin (says The Times) will be for the transmitting apparatus, and the second (North Petherton) for the receiving end of the station, which is to be constructed by the Marconi Company and to be ready for work within six months from the date of the provision of the site by the Post Office. This station is intended to be part of the scheme for ensuring wireless communication with the Dominions, and it is provided that similar stations shall eventually be erected in India, Canada, Australia and South Africa. At present the Canadian station is the most forward; some parts of the work are almost completed, and the masts have been built. There are two such stations in Canada – one at Montreal to work with England, and the other at Vancouver to work with Australia. There will ultimately be four other stations on this side to communicate with each of the Dominions named. The South African station is under construction at Cape Town; the Australian Company has accepted the Marconi Company's tender for the construction of a station but the site has not yet been fixed; and India has also agreed on the terms of its contract so that the preliminary contracts for the chain of Empire Beam Wireless Stations are now complete. The Marconi Company is also at work on the construction of its own Beam Wireless Station for communicating with America and the Continent at a site two miles from Dorchester. In each case the wavelength is below 100 metres. The height of the aerial mast and the length of the aerial are dependent on the wavelength and the masts will be about 300 feet and about 650 feet apart.

The Beam Wireless is understood to be proving for efficacious and cheaper than the broadcast. The wave is concentrated directionally upon the station for which it is intended and the result is not only greater reliability for the message but also considerable economy in transmission.

We understand that negotiations for the purchase of the property referred to have now been completed, and that in addition two more fields will be required in connection with the scheme.

A special meeting of the North Petherton Parish Council was held on Tuesday evening to consider the footpath question arising out of the proposal to purchase Copse Farm. Mr. John Slocombe J.P. was in the chair, and he read a letter from the General Post Office London, asking whether there was any likelihood of the closing of the footpath being objected to.

It was pointed out that the path was not a public one, but a convenience to approach Park Farm and some members mentioned that they felt assured no objection was likely to come from that direction. The Council unanimously agreed to reply to the Postmaster General that no objection would be likely to arise from the closing of the path.

In June 1925, the process of purchasing the required land by the Post Office continued; however, there appeared to be some friction between the Post Office and Bridgwater Rural District Council with regard to staff accommodation. The events were duly recorded in the local press that month:

North Petherton Wireless Station

The Clerk reported in regard to the application from the Post Office authorities for power to close a footpath from

Huntworth Lane to Petherton Park, in connection with the proposed Wireless Station at North Petherton, that the Parish Council had passed the necessary resolution agreeing to the stoppage of such path and a further resolution by the Parish Council that no expenses should be borne by them in connection with the stoppage of the path. The District Council also decided to pass similar resolutions. A letter was read from the Postmaster General stating that in connection with the establishment it was desirable that four houses should be available for the use of the married staff to be employed but, unfortunately no such accommodation was apparently to be had at present on rental terms.

It was understood however that the Bridgwater Rural District Council were about to erect some houses at North Petherton and the Postmaster General asked whether the council would be good enough to earmark four of these houses for the use of the department's staff.

Mr. West: "Nothing doing!"

The Clerk pointed out that they had not yet had applications for all the ten houses to be erected at North Petherton. Mr. Slocombe thought it was too early yet to assume that all the houses would not be applied for. People were afraid to send in applications because they did not know when the houses would be started.

The Clerk: "Or what the rents will be. I suggest you file this letter in the ordinary way with the rest of the application."

The Chairman said he supposed when the time came this application would be considered with any others. He took it there was no obligation on the part of the Council to provide houses for Government officials.

The Clerk's suggestion was adopted.

Despite the above, progress of the installation continued unabated, and a progress report duly appeared in the local press:

The North Petherton Station – Progress of the work

For some months now, there has been considerable surmise as to the exact nature of the new Wireless Station in course of construction at Huntworth, North Petherton. With the kind permission of Colonel T. F. Purves M.I.E.E., Engineer-in-Chief of the General Post Office (Radio Section) London, the local representative has had the pleasure of going over the site with the Officer-in-Charge Mr. T. G. Kennard and was greatly interested in the description of the construction and the subsequent working of the station.

As is generally known, about 300 acres of land situated between Huntworth Lane and the private road leading to Park House was recently purchased by the Post Office authorities from Lord Portman and Mr. F. Hine.

An entrance has been effected from Huntworth Lane and this has necessitated the removal of a high bank and the construction of a roadway in which over 100 tons of stone were used. This leads to the 'dump' where stores are kept and a giant concrete mixer has been erected. The 'dump' is connected with various parts of the site by a network of light railways approximately three miles in length. The station will be comprised of two aerial systems, one of which will connect with a sending station in Canada and the other with South Africa. A feature of this erection of the aerials is that they must be exactly at right angles to similar stations at present in course of erection in Canada and South Africa.

Masts 290 feet high

The aerial system at Huntworth will be supported by ten steel lattice masts. Each mast will be erected on four concrete bases each having capacities of 384 cubic feet respectively, each mast will be 290 feet high and 12 feet square at the base. These masts, erected 650 feet apart and built on the cantilever principle are practically self-supporting but in order to provide against wind strain etc. each will be supported by four stays or cables, a stay foundation comprising approximately 504 cubic feet of solid concrete.

Equally-balanced 15 feet from the top of each mast will be a cross arm support 90 feet in width. Mast and cross arms have an approximate deadweight of between 45 and 50 tons. The concrete alone for each mast foundation will weigh almost 93 tons and the four blocks for stay foundations will require 120 tons of concrete.

The aerial system

The system is quite unlike the present commercial system of wireless, being known as the 'beam' which means that a beam of wireless waves are transmitted from the sending station and having a directional effect can be received practically only by the station intended. The system is thus practically secret.

The aerials will be suspended perpendicularly between the masts from each of the balanced cross arms, and will form a kind of screen reaching within a few feet of the ground. Ninety feet behind these will be what is known as the reflecting screen which be almost similar in construction and have the effect of collecting the waves which are not gathered by the front screen. The total length of each screen will be 650 feet. Current will be obtained from a plant to be erected on the site midway between the masts.

The Huntworth station is being constructed entirely for receiving the transmitting station in connections with the system being at present in the course of construction at Bodmin, Cornwall.

The work is being carried out by the Marconi Company under the direction of the General Post Office and we understand that its perfection is due to the many years of stringent study already given to subject by Signor Marconi.

It is a matter of interest and satisfaction to learn that the stone, sand, timber etc. used in the construction of the station is being purchased locally, and that approximately 105 men are at present employed there, and with few exceptions they are all local residents.

It will also be a source of comfort to local wireless enthusiasts to learn that the proximity of the station is not expected to interfere in any way with receiving sets in the vicinity. When completed this will be the first radio beam station under the direction of the General Post Office."

There is little doubt that the construction of the station was of great interest to the local inhabitants of the area; wireless communication was very much in its infancy, and there was surely a great deal of wonderment that these large aerial systems could enable contact with countries previously only seen on maps or in atlases.

A sense of great mystery prevailed for many years, as the locals got to grips with the magnitude of what was happening just outside their village.

A report in *'Wireless World'* regarding the actual erection of the masts from October 1925 reveals how complex the operation was:

"The site for this station has been selected with great care. For strategic purposes it was thought advisable that the stations should be well inland, so that the probability of bombardment from enemy warships, in the event of war, should be reduced to a minimum. It was also desirable that the station should stand on high ground with a clear stretch between it and the distant sister station with freedom from hills and trees in the vicinity.

The station buildings under construction.

The method employed in the erection of the masts is that as soon as the foundations are ready, some of the lower sections and braces are set up and bolted into a vertical position. A derrick is then moved into position with its base within the four corners of the tower, but with its weight resting in such a manner that the anchorage forms a hinge.

A block-and-tackle is made fast to a point about two-thirds of the way up the derrick, and by this means the derrick is hoisted up from the ground. A working platform is then assembled inside the mast and surrounding the derrick.

With the lower sections, platform and derrick all in position, the steelwork for the next section of the mast is hoisted into position piece by piece, and bolted into place. When this is completed, the whole derrick is raised up from the ground vertically, and made fast to the lower portion of the mast. More steelwork is built round it; then the derrick is again raised, so that eventually the derrick stands with its head projecting above the finished walls of the tower.

When the final stage is reached, auxiliary derricks are hoisted into place, and then the four great portions which form the cross arms, are raised and bolted into place. Finally, the derricks are dismantled and lowered to the ground."

The process of erecting the aerials was dangerous work, and one rigger from Bridgwater was killed whilst erecting a similar aerial at the Tetney Beam Station in late 1925. The inquest into his death reveals just how dangerous the process was, and is worth reproducing to emphasise the scant regard to Health & Safety procedures which we now regard as normal:

One of the earliest general views of the
Bridgwater aerial system.

A cross arm in four sections ready to be
bolted to the masthead.

"The terrible death of a workman employed in the erecting of one of the steel masts at the Empire Beam Wireless transmitting station, which the Marconi Company are building at Tetney, formed the subject of an inquiry by the Louth District Coroner (Mr. Herbert Sharpley) at the station on Saturday.

There are several of these on the station, and the mast from which the workman fell a distance of 190 or 200 feet is 277 feet in total height.

The deceased was a steel erector named Albert George Russell, of 6 Price's Buildings, Salmon Parade, Bridgwater, in the employ of Messrs. Francis Morton and Son. Ltd., of Hamilton Ironworks, Garston, Liverpool, who are the sub-contractors building the mast, and he had been lodging in Tetney during the time he had been engaged in this work.

Thomas C. Percy identified the body as that of his brother-in-law, who he said was 38 years of age, and had a wife living at Bridgwater.

Percy Hagger, charge hand, stated that he lived at present in Grimsby, and was in employ of Messrs. Morton and Co. Ltd., as a sort of loading hand. He was in charge on Thursday morning. He was working there the early part of the morning, but they came down to breakfast, and returned a little later after nine.

Witness was one stage above the deceased. They fixed the brace and then fixed the ladder. The height of the mast was 277 feet, and where they were working was about 190 or 200 feet from the ground. Deceased had been working with witness above five weeks. On this particular morning they worked along in the ordinary way, and then suddenly witness turned round - he did not know why - and then

151

saw Russell falling. Witness shouted to the foreman down below. He could not say how Russell came to fall. He had been with witness on ten other masts, and had never slipped before. He was a most careful and steady man. They were all working inside, but he fell on the outside because he was approaching another point. He was going up one of the diagonals to approach the other point, and somehow slipped and came down outside.

Witness had partly engaged the deceased at Bridgwater and prior coming here deceased had been accustomed to the kind of work he was doing there. There were ten exactly similar masts built at Bridgwater. Russell had been doing the topping masts at Tetney, but Thursday morning was the first time he had gone up this particular mast, and witness gave him instructions. "Topping" was putting the cross-head on. Russell started from the bottom and worked his way up, and was engaged on the joint above the ladder. The men were putting in the bracings to take the ladder. Deceased had been engaged in bolting the bracing.

There were three other men working the same mast, and one named Hagger was 13ft. 6in. higher up the mast. Between ten and eleven o'clock witness saw Russell falling, when he had fallen to about 15ft. from the ground. Witness was standing on one side of the mast, and Russell fell on the other. Hagger called out, and witness looked up and saw the deceased at the distance he had stated.

There would be no means of fastening one's self while engaged in this work, because one would be moving every few minutes. There were the angles to hold on all the while. It was not so cold on the morning of the inquest - the sun was shining - and witness did not think that the deceased's hands were numbed. He had not the slightest idea how the accident occurred. Russell fell clear of the mast outside, but his work was inside, on the internal braces. He must have come through one of the openings, and, having fallen through, there would be nothing to enable him to save himself. If the men used fasteners, they would be fastening themselves all the time and get no work done.

Witness went to Russell, but the man did not say a word - he just gave one breath and then seemed to go. As far as witness could judge, he was dead as soon as he had reached the ground. He fell on his left side. Since they had been at Tetney they not had any accident before this, and it was very seldom they had any. At Bridgwater there was not a single accident - not a little skin on a man's hands. Deceased was a steady man.

In falling he would not strike anything. He simply fell down with a rotary movement.

P.C. Haynes said that when he arrived the man was quite dead. Dr. Frank Herbert Rotherham, of Grimsby, said he was sent for on Thursday when the accident happened, and came at once. The man had been removed from the foot of the mast to the shed, and witness examined him. He was quite dead.

Witness thought death had been instantaneous from the injuries. Russell had a fractured spine, fracture of the base the skull, and fracture of both thighs. Mr. Watkins said the deceased had been doing exactly the same work on the whole of the ten masts at Bridgwater. It was risky work -

the men had never to forget for a moment that they were standing on a bar three inches wide. That was the danger in their getting accustomed to it. It was rather a curious feature that Russell did not shout out at all - neither Hagger nor Smith heard him shout.

The Coroner said it seemed him entirely an accident, and there was nobody who could explain how it happened - only the deceased could have said how he came to slip. He should return a verdict to the effect that he died from the serious injuries described by Dr. Rotherham, and that he received them accidentally falling from the mast from which he was at work."

The matter of the Beam Wireless Network was debated in Parliament on 20th July 1925, and the Postmaster General Sir W. Mitchell-Thomson gave an update on the Bridgwater station during one of his speeches regarding the status of the network:

"The company, however, subsequently came to the conclusion that for technical reasons the original scheme of concentrating all the four sending stations on one site and the four receiving stations on the other site was impracticable, and, accordingly, different arrangements were made, find another agreement was entered into providing for the erection of two groups of two stations each, one group in the South West of England, at Bodmin and at Bridgwater, for communication with Canada and South Africa, and the other group on the East Coast for communication with India and Australia. There were considerable difficulties, technical and other, in finding suitable sites.

The mast being hoisted into position.

But eventually sites were found at Bodmin and Bridgwater. They were handed over to the company on the 6th April, and the company tell me that considerable progress has been made in connection with the erection of the masts and buildings at Bodmin, and that work is proceeding on the excavations and foundations for the masts at Bridgwater. The stations are due to be completed, under the terms of the contract, by the 6th October. The sites for the sending and receiving stations for India and Australia also presented considerable technical difficulties, but they have been finally settled at Grimsby and Skegness.

The necessary legal formalities are just being completed for their transfer, and I have actually placed the order for the stations with the Marconi Company in anticipation of the completion of the legal formalities. Under this contract the stations are to be completed within nine months."

The Post Office were very keen to publicise the new station and many reporters, representing both the local and national press, were invited to the site of the Bridgwater station and to publish articles reflecting the 'wonder of wireless communication'. An excellent example of this appeared in the 14th September 1925 issue of the *'Western Daily Press'*, along with many photographs of the site:

GOVERNMENT WIRELESS DEVELOPMENTS

EXCLUSIVE PICTURES OF BRIDGWATER "BEAM" STATION

The accompanying series of photographs, the first to be published of the new 'beam' station being erected near Bridgwater by the General Post Office, for working on short wave radio-telegraphy between this country and Canada, and between here and South Africa. Motorists travelling along the main Bristol-Bridgwater Road have noticed with interest and not a little curiosity, the appearance away on the left, of several immensely tall lattice-work masts. Their position is some distance removed from the main road, and speculation has been rife as to the precise object of these giant landmarks. One our representatives, accompanied by a *'Western Daily Press'* photographer, had the opportunity, the other day, through the courtesy of the Engineer-in-Chief of the General Post Office, Col. T. F. Purves, O.B E., M.I.E.E., of making a tour of the latest wireless station, and of gleaning much information which should interest the public general, and wireless enthusiasts particularly.

The foundations at the base of one of the masts.

The site is situated at Huntworth, North Petherton, but it is officially designated the Bridgwater Beam Station. Some time ago the Post Office authorities, having searched for a suitable position upon which to erect this first station of its picked upon about 300 acres of grass and partly wooded land, owned by Lord Portman and Mr. Hine, the well-known farmer and agriculturalist. The whole area was purchased outright by the Government, and in April last a commencement was made with the work of construction. Natural obstacles such as banks, trees, etc., had to be overcome in laying out the site, and there were certain difficulties to be encountered through the absence of sun in taking sightings. However, under the direction of Mr. T. G. Kennard, the Engineer-in-Charge from the G.P.O., excellent progress has been made, and at the present moment six of the 10 great masts which will play so important a part the ultimate scheme, have been and a seventh is rapidly growing in stature.

The First Bearings Taken

Five masts are necessary to each station – the Canadian and the South African. Mr Kennard escorted our representatives from the entrance to the site, and the vastness of the area soon became apparent from the long tramps from one point to another. A certain spot attention was directed to a wooden peg driven into the ground. It was from here that the first bearings were taken for deciding the actual positions of the masts. Bridgwater is to be a receiving station, transmission being carried on at Bodmin, Cornwall, at a station to be erected on similar lines. Viewing the completed masts from the spot at which the original bearings were taken, it is seen that the lofty erections form the shape of a 'V', one leg of which is at right angles to Canada, and the other to South Africa.

Auxiliary derricks employed to fix the cross arm into position at the top of each mast.

Some idea of the immense proportions of the masts may be gained from measurements supplied to the writer by Mr Kennard. Each mast is 287 feet high, and 12 feet square, being built on the lattice-work principle. To each mast is a set of four supporting stays. At the top is a cross-arm, 90 feet long, which will be at right angles to the aerial.

The object of this is to suspend the aerial on one side, and to provide a reflector on the other. Thus, the aerials will differ from the better known type. They will hang from the cross-arms to within a foot or two of the ground, being kept in position by dead weights, and upon them will be directed the 'Beam', from the sending station; a reflecting screen to be situated to the rear of the aerial, and similarly constructed, will collect and reflect waves not already gathered.

Remarkable Figures

The masts are spaced at 65 feet from each other. They are so constructed as to stand unaided, but, as a matter of fact, every one of them is erected upon foundations which, from appearances, would seem to have been fashioned to remain for all time. These foundations consist each of 384 cubic feet capacity, and are of solid concrete.

The stay anchors are each held down by a concrete block of 504 cubic feet capacity, and here, as in the case of the mast foundations, the amount of concrete used may be estimated as 135 pounds per cubic foot. Each mast weighs approximately 50 tons, and each cross-arm at the top, seven tons. The aerials will be raised by hand winches, each of five tons capacity.

Figures and other such indications, however, cannot impress one's mind with the colossal nature of the undertaking. It is necessary to stand at the foot of one of these sky-scraping towers and gaze upwards, to get a real idea of its loftiness. The Clifton Suspension Bridge is 245 feet above the Avon at high tide, 40 feet less than the height of these amazing structures! Running up the inside of each mast is an iron ladder, erected as the mast is, piece by piece. The rungs are 10 inches apart, so that there are about 344 rungs to negotiate to reach the top!

The station will be absolutely self-contained, that is to say, within the red-bricked building which is being brought to completion on the site, and will be the receiving plant and the battery-charging and electric lighting apparatus. When operations commence after the station has been established, there will be a comparatively small staff of highly skilled men put in possession, and it is understood that the Post Office will build houses for them upon the site.

For the purposes of men employed since the acquisition of the area, and on account of the distances to be traversed every day getting from one end of the site to another, a miniature railway was laid down, and about three miles of track have been in constant use. Altogether 175 hands have been employed on the work of preparing the ground, and latterly in raising the masts, cross-arms, etc. The iron latticework was travelled from Scotland by rail, but all the cement, sand, and stone and timber have been obtained from local contractors, the stone having been locally quarried.

The derrick in position.

The great majority of the workmen have also been obtained locally. In the course of clearing the site it was found necessary to fell a number of big trees, but care has been exercised to restrict this part of the operations as far as possible. One of our photographs depicts a number of the men perched in various perilous positions upon a partly erected mast. Every one of them is a skilled worker, and the dauntless manner which they climb and turn and twist on the ironwork, at all heights and angles, is equal to many a first-rate trapeze artiste's turn.

First of its Kind

The Bridgwater Station is to be a commercial station, and there can be little doubt that, ultimately its establishment will have a beneficial effect so far as the general public is concerned, in the gradual cheapening of rates and a speeding up of the telegraphic service. It is the first station of its kind to be used by the Post Office. Many amateur wireless devotees will probably be wondering, particularly those residing in the immediate vicinity of Bridgwater, what, if any, effect this beam station will have upon their own activities. The writer was assured that there need be no concern in this respect, as it is not anticipated that the proximity of the station will in any way interfere with receiving sets.

It is expected that the whole of the work will be completed shortly, and communication will at once be established. Thus another immensely important step forward will have been taken in the most wonderful of modern sciences. The Post Office authorities regard the site they have acquired for the setting up of the station as an ideal one, and results are expected to of the most satisfactory character.

The station will be operated from London by means of telegraph lines connecting up with the new loaded main cable recently laid underground between Bridgwater and Taunton."

The *'Western Daily Press'* in its issue of 8th January 1926 was very keen to extol the virtues of the new station, as it excitedly reported:

"The year just begun bids fair to be one the most important in the history of commercial wireless telegraphy as it affects Great Britain and the Empire. Before many months have passed, the first of the new Marconi-type beam stations will be brought into operation, and in the course of the year direct high-speed wireless services on the short wave beam system will be established with all the principal Dominions.

The first of these beam stations - for communication with Canada and South Africa - are at an advanced state of construction. All the masts have been erected, the buildings completed, and the machinery and some of the wireless apparatus installed. The main part of the wireless apparatus - the transmitters and receivers - is now being tested, and it will be installed as soon as these tests are completed. Meanwhile, work is progressing on the erection of the aerial and feeder systems, and fencing and road-making.

The stations now nearing completion are at Bridgwater, Somerset, and at Bodmin, in Cornwall. The Bodmin station will be the transmitting station used for communication with Canada and South Africa, and the Bridgwater station will be the receiving station for these two services. At each station there are ten masts, five for communication with each of the two Dominions.

The design of the masts is identical for the transmitting and receiving stations. In the one case the aerial and reflector are utilised to concentrate the wireless energy and transmit signals in a particular direction, and in the other case the same arrangement is used to collect the received energy and concentrate it, thereby still further increasing the signal strength.

Five masts for each Dominion are erected in a straight line at right angles to the direction in which communication is to be established. These masts are 277 feet high, each having a cross arm at the top, measuring 90 feet from end to end. The aerials and reflectors will consist of a number of vertical wires suspended from triatics attached to the cross arms of the masts.

There will thus be two parallel steel cables, separated by a distance dependent on the wavelength used, running on each side of the masts from the first to the last. From these cables, the vertical aerial and reflector wires will be suspended, the lower ends of the wires being kept in position by balance weights. The distance between the masts is 650 feet, from centre to centre, and the length of the whole system of five masts for each transmitter is about 3,200 feet from tail anchor block to tail anchor block.

At the Bridgwater receiving station, power is supplied by 18 h.p. two-cylinder Aster engines driving the d.c. generators which supply the station with light and run the motors for charging the receiver batteries. The buildings here are smaller than those at the transmitting station, owing to the smaller amount of power required. Both transmitting and receiving stations will be connected by direct land lines with the Central Telegraph Office, G.P.O. London, and the transmitting station will be operated from London by distant control.

The incoming signals will be automatically relayed to the land line at Bridgwater, and passed on to the Telegraph Office, so that both stations will be controlled from London. The outgoing and incoming messages will be dealt with at the same table in the telegraph office thus giving true duplex working and complete central control over the traffic.

The corresponding stations in Canada, near Montreal, and South Africa, near Cape Town, are in practically the same state of advancement as the English stations, and other similar stations in the Imperial system are being erected at Grimsby and Skegness, England, for communication with other stations at Poona, in India, and Melbourne, in Australia."

On 20th October 1926, Guglielmo Marconi himself gave a speech to the press with regard to the opening of the new Bridgwater receiving station. It is clear that Marconi was a visionary in that all his predictions did indeed come to pass, and for historical reasons alone it is worth reproducing the text in full:

"In addressing the representatives of the press, I have the satisfaction of feeling that I am speaking to persons who already know a good deal about telegraph services and who can fully appreciate the importance of speed and accuracy in the transmission of messages.

I am also glad to meet you at a time when the whole theory and practice of long distance radio communication is undergoing a most radical and beneficial change what, in fact we might be justified in terming, a revolution.

I am glad to be able to tell you that on Monday morning we received from the Engineer-in-Chief of the Post Office the following official certificate (see also Appendix 3):

Office of the Engineer in Chief
General Post Office (Alder House)
LONDON E.C.1.
18th October 1926

Beam Agreement dated 28th July 1924
The Preliminary Certificate

This is to certify that the sending Beam Station erected at Bodmin for transmission to Canada and the receiving Beam Station erected at Bridgwater for reception from Canada have been submitted to seven consecutive days' working and after making due allowance for the period during which the abnormal electric storm was experienced, the stations satisfactorily fulfilled the conditions of being capable of sending and receiving at the same time to and from the Canadian Station, one hundred words of five letters each per minute during a daily average of eighteen hours in accordance with Clause 6 of the Beam Agreement.

(Signed) T. F. Purves
Engineer-in-Chief

The Post Office inform us that they have taken over the stations and announce that the high-speed wireless telegraph service between Great Britain and Canada through these stations will be opened to the public on Sunday midnight the 24th instant.

What is the Beam System?

The principal characteristic of the new system is that the waves which carry the messages are transmitted or projected within a small angle in any desired direction instead of being allowed, as has been done up to now, to spread or be broadcasted around in all directions. This limitation of the propagation of the waves to a given direction bring about a strengthening, or concentration, of the energy on the station with which it is desired to communicate, with the result that the electrical power of the sending station is utilised very much more efficiently and to a degree enormously in excess of what was hitherto believed to be attainable.

The increased efficiency makes it possible for stations utilising only 20 kW or about the equivalent of 26.8 horse power of electrical energy in the sending aerial, to give much better service across the Atlantic than stations on the older system of long waves employing hundreds or even thousands of kilowatts.

A further and considerable advantage is brought about by the utilisation of short waves as it is only with these waves, in consequence of their far greater frequency of oscillation, that it is possible to work telegraphic services at really high speeds, whilst speeds of the same order have been quite attainable with the long waves which up to now have been in general use for long-distance radio communication.

But short waves used in the ordinary way would not allow of these high speeds being attained over long distance unless an enormous amount of electrical energy could be employed, and I will tell you the reason why.

The tests between England and Canada have already shown that the use of Beam Aerials and reflectors at both ends has resulted in a signal strength some 100 times what obtainable with non-directional transmitting and receiving aerials utilising the same power.

Now, it is easy to calculate that in order to receive signals 100 times the strength it is necessary to use 10,000 times the energy at the transmitting end, and therefore, as the power supplied to the transmitting aerial of the Beam Station is 20 kW. 20,000 kW, which is an impossible and absurd amount of energy, would have to be used instead of 20 kW to give the same average strength of signals at the receiving station. Moreover, the use of Beam Aerials has enormously reduced the amount of interference from atmospheric disturbances. I think we all know that 'atmospherics' have been the bugbear of wireless, but with the introduction of this new type of station, I am satisfied that they have ceased to exist as a serious hindrance to the working of high-speed radio communication between England and Canada. I fully realise that this is a bold statement, but I feel pretty confident that I am right in making it.

Fading, or a frequent attenuation, of signal strength has been a marked feature of wireless transmission over long distances, especially when short waves were employed, and although in my experience fading appears to be worse with wavelengths between 200 and 1000 metres, it is sometimes experienced on the 26 metre wave which has been utilised by the Beam Stations for communicating between England and Canada. According to the experience of myself and my collaborators the use of reflectors has the advantage, which is perhaps common to all sharply directional systems, of diminishing fading. This is brought about partly no doubt by the enormous increase of the received signal strength which is obtained by the utilisation of the Beam System and which thereby increases the margin of readability of the received signals, but fading certainly does still exist on the Bodmin-Canada circuit, and on one or two occasions has cut off all communication between the two stations. Two such bad periods occurred on the 20th September and on the 14th October.

Curiously though, these bad periods coincided with the appearance of very large sunspots and intense Aurorae Borealis in Canada and at the same time the telegraph lines and cables were thrown out of action or the working of them was greatly interfered with. We noticed, however, during these bad periods that reliable signals could be obtained across the Atlantic by using another wave, not of 26 metres, and as provision has been made for using two waves at the Beam Stations. This additional wave will be employed during the exceptional conditions which may interfere with the transmission of messages on the 26-metre wave.

The official Post Office requirements laid down that the stations for the Canadian service should be capable of communicating at a speed of 500 letters per minute each way (exclusive of any repetitions necessary to ensure accuracy) during a daily average of 18 hours and that a demonstration fulfilling these conditions should be given by actual working to and from Canada for seven consecutive days. This test took place between October 7th and October 14th and the guarantees, which were regarded by everyone as being extremely stringent, have been successfully fulfilled. During these and the preliminary tests, speeds of 1,250 letters per minute in each direction equal to 2,500 letters per minute over the whole circuit have been worked for hours on end. Counting every hour of the 7-day test, the average speed of signals has been about 600 letters per minute in each direction or 1,200 letters per minute over the complete circuit.

Operating has been carried out from the Central Radio Office at the General Post Office London, from and to the Telegraph Office of the Canadian Marconi Company in Montreal.

These results have abundantly justified our faith in this new system and indicate that the stations will be more than capable of handling all the traffic that is likely to be available between England and Canada for some years to come.

I have been working systematically on the Beam System for over ten years, and during nearly all this time have been valuably assisted by Mr. C. S. Franklin, who has followed up the subject with great thoroughness and to whom a great part of the credit for our present success is due. As long ago as June 1923, following the results I obtained on a cruise of the 'S. Y. Elettra', I had already made up my mind that this new system was going to carry a considerable part, perhaps of all the most important high-speed long distance traffic of the world, and I stated so in my report submitted to the Directors of the Marconi Company on June 23rd 1923, and also in a paper which I read before the Royal Society of Arts on the 2nd July 1924. I feel confident that my belief is going to be justified by the results, not only for communicating with Canada, but also in regard to South Africa, India and Australia, as soon as the stations intended for these services, which are in an advanced state of construction, have been completed.

On the 13th October, during the official tests, the following message was received from the Canadian Company:

> This message transmitted at two hundred and fifty words per minute demonstrated the possibilities of speedy and accurate inter-Empire communication now made possible by radio by means of Marconi's Beam System stop High speed of transmission in conjunction with direct communication annihilates distance and bring the members of the British Empire in such close touch with each other that one can almost visualise delegates to

Empire Conferences in future representing their countries without leaving their own firesides and yet enjoying all the advantages that personal contact offers stop Dated at Montreal thirteenth A/D one thousand and nine hundred and twenty six ends.

I have here the actual tapes on which this message was recorded at a speed of 250 words per minute. I do not think that there is any other long distance telegraph service in the world which could show anything like so great a speed.

I would like to add that I do not think the Beam System is by any means limited to wireless telegraphy. I feel confident that it can be utilised for placing wireless telephony on a much more practical basis than it is at present, besides helping the system of picture and facsimile transmission, not to speak of television.

Even for what is called broadcasting, I believe it will result in enabling programmes and speeches to be transmitted to large portions of the United States, Canada, South Africa and Australia with much greater strength and accuracy than it is possible to obtain by means of the existing broadcasting system.

I wish again before concluding to say how much the perfecting and establishment of the Beam System owes to the skill and perseverance of the engineers and experts of the Company and particularly to Mr. C. S. Franklin on the technical and scientific side and the Company's Engineer-in-Chief, Mr. R. N. Vyvyan, on the engineering side.

I also wish to express my appreciation of the consideration and encouragement we have invariably received from the authorities of the Post Office and of the cordiality and good feeling which has existed between the Post Office engineers and our staff during all the time the stations have been under construction and also during the tests which have lately been carried out.

I feel confident that this new system will be efficiently worked at this end by the Post Office to the advantage of all users.

Marconi staff who conducted the original Beam tests at Bridgwater, prior to handing over for Post Office tests, October 1926. L-R: J.A. Smale, H.M. Copland, W.J. Cotter, H.S. Gibbs, C.S. Franklin, L. Kemp, J.G. Robb, and M.B. Hunter.

The regional press also reported this momentous occasion on 22nd October 1926:

"The Postmaster General announces that the Canadian Beam Service, the first of the new Imperial Wireless services, will be opened on Monday morning next. The new service, which supersedes the former Marconi Wireless service between London and Louisburg, Nova Scotia), is operated on an improved system directly between the Central Telegraph Office, London, and Montreal. It was demonstrated to journalists at the General Post Office yesterday, when Mr John Lee, Controller of the Central Telegraph Office, explained its workings.

The first message sent was one of greetings from the London Press Representatives to Montreal, to their colleagues throughout the Empire. It was despatched at 4.22 at a speed of over 100 words a minute, and two and half minutes later an 'O.K.' signal was received from Montreal, indicating that it had been received.

Later a reply in French came through from Montreal, where Canadian Pressmen were watching the reception from London. Mr Lee said the Beam service known as *'Empiradio'* was to be used for the present in co-ordination with the cable service, and the public were to be given the opportunity of marking telegrams with an alternative routing thus providing against the choking of the service. The charges were ordinary rate, 9d a word, deferred rate 4d a word, night letter telegram 4s 6d for twenty words. A new class of traffic, to be known as post letter telegrams was to be introduced on the Beam service at 1½d a word, with a minimum of 2s 6d for 20 words or less. The messages would be collected and delivered by post at each end.

Personal messages of congratulation and appreciation were passed yesterday by the Beam Wireless to Lord Atholstan, Leader of the Canadian Press at Montreal, and Lord Burnham, President of the Imperial Press Conference. Future developments of the Beam service will be to South Africa, India, and Australia. Signor Mussolini on behalf of the Italian Government and people, yesterday sent a congratulatory message to Commandatore Marconi on the inauguration of his discovery of the wireless beam service to Canada."

The station officially opened on 25th October 1926, although the first day of operation didn't quite go to plan, according to reports published on 27th October:

"Regarding Monday's breakdown of the new Beam Wireless service to Canada, said to have been due to atmospherics, Senatore Marconi, in a statement last night, declared that the fading experienced had been quite wrongly attributed to that cause. It had not yet been possible to ascertain what cause was attributable, and possibly some insulation fault was responsible. It was obvious, he added, that an entirely new system could not be expected to be perfect from the very start."

Despite the initial problems, the local press were keen to use the station as excellent publicity for the town of Bridgwater itself:

Beam Wireless

The inauguration of the 'beam' system marks another advance in wireless. Incidentally it has given Bridgwater a useful 'boost' for although the installation in the country is at North Petherton, Bridgwater as the nearest town had had all the publicity advantages attending the new development. The system starts working with Canada but is intended to link up the Empire by Beam Wireless. Briefly, the feature of the Beam System is that wireless energy is concentrated into a directional beam as the car lamp focuses light. The beam passes straight from one station to another and cannot be tapped by listeners-in except those on the beam. A short 'wave' is used, and this has not only high speed (about 600 words a minute in each direction), but is comparatively free from atmospheric interruption.

The system between this country and Canada was inaugurated last week. At the outset it has not worked with perfect smoothness, thought there is no doubt of its ultimate success, compared with Rugby, which works at high power on a long wave, the beam installation has cost less than one-tenth. Marconi predicts that the beam principle will be extended to wireless telephony, to the transmission of pictures, the broadcasting of speeches, and even to television. So many of the fairy tales of science we have seen realised who will doubt that these tools will become the accomplished actualities of a perhaps not too distant future.

The existing and powerful Post Office long-distance transmitting station at Rugby did not seem overly concerned about the new service, despite some worries that the station might be superseded by the new Beam Wireless service. A local newspaper in Rugby carried an interview with the station manager at Rugby, who sought to reassure staff and customers alike:

"The inauguration of the beam system of wireless telegraphy from Bodmin to Canada has again brought forward the question of this method ultimately superseding a super wireless station such as Rugby, on the grounds of economy and speed. Although this idea has been conveyed in recently published statements, it seems to have overlooked that Rugby Radio Station fulfils a particular function not encroached upon by the newer method.

"Bodmin would be unable to do the traffic we are now doing," Mr. H. Faulkner, engineer-in-charge of the Rugby Station said in an interview to representative of the *'Rugby Advertiser'*. Bodmin is designed to work to one place and most of our traffic involves working to the whole world. Our Foreign Office press is received by a great number of stations in all the Colonies and dependencies and ships in all parts of the world with the one transmission."

"Another service we do is known as long distance radio. One hands in message at a Post Office and it can be sent to any station in the world fitted to receive it. Obviously the beam system would not be capable of giving that kind of service. For a beam station there must be a separate aerial system for each service. For working from one particular place to another particular place it has obviously many advantages, but, also obviously, it is of no use for broadcasting."

"There is no reason why a broadcasting service which will reach all parts of the world at the same instant should not have a big use, and be cheaper for a universal press service than using separate transmitting stations for each direction, although the latter costs less individually than one big station."

"There no doubt that in case of war this station would have very great strategic advantages as far as the Navy is concerned by reaching ships in any part of the world. The beam system would not be able to perform that function, as far can be seen at present."

With the initial work completed, the Bridgwater station provided a valuable and effective receiving link for many years.

CHAPTER 5

OPERATION OF THE BRIDGWATER
BEAM WIRELESS STATION

It will be useful to describe the workings of the station, together with details of the equipment used. The operation of the Bridgwater station would have been similar to the other receiving stations in the beam network, although of course it did have some local 'peculiarities'.

A detailed explanation of the operation appeared in an article entitled 'The Wireless Beam' which appeared in *'Wireless World'* on 24th November 1926. The text, although somewhat technical, emphasises the advanced radio circuitry of the time. The article is reproduced below, with permission.

"The Bridgwater Beam Receiving Station is built on land situated near the village of North Petherton, 2½ miles south of Bridgwater, Somerset, the site being three-quarters of a mile from the main Bridgwater-Taunton road. It is about 60 feet above sea level, with open country around in all directions, so there is no screening to interfere with the reception of signals.

The buildings are of brick, and are divided into two main sections; the engine room and the receiver room and offices – joined by a short covered passage. The continuity between these sections is definitely broken, to avoid the transmission of vibration from the engines to the receivers.

Power is generated by two 18 h.p. Aster two-cylinder petrol-paraffin engines direct coupled to two Metropolitan-Vickers 10 kilowatt dynamos supplying power at 110 volts to the main switchboard, for charging the battery used for lighting the station, and for driving four motor generator sets manufactured by the Crypto Electrical Company.

Two of these sets are for charging the four-cell filament batteries and two for charging the 110-volt anode batteries. The plant is thus duplicated throughout.

Adjoining the engine room are the station stores and station battery room. The station battery is a 60-cell battery by the Tudor Accumulator Co. of 132 ampere-hour capacity.

The filament and anode battery room contains two 760 ampere-hour four-cell filament batteries and two 30 ampere-hour 110-volt anode batteries by the Tudor Accumulator Company.

The receiver room is at the west end of the block. The receivers stand near the west wall where the feeders enter the building. At right angles to the receivers is a table for the recording apparatus and land-line instruments. The received signals are put direct to line, through A.T.M. high-speed relays, and, when required, can be checked for formation by means of undulators. A Wheatstone transmitter is also installed for testing purposes.

A Post Office omnibus sounder circuit connects the Central Radio Office in London with the Bridgwater and Bodmin stations for control purposes, and there are, in addition, two direct telegraph lines to London for the transmission of received signals to the C.R.O.

The Bridgwater receivers are arranged for reception of signals from Canada on a wavelength of just over 26 metres. As an alternative, the receivers may receive signals from Canada using the other aerial bay. The call letters of the Canadian transmitting station are CG.

The masts, receiving aerial and feeder system are duplicates of those at the Bodmin beam transmitting station, except that the feeder tubes at the receiving station are slightly smaller than those at the transmitting station. The receiving masts are erected at right angles to the direction from which the signals are to be received, their exact orientation being:

Canadian line of masts: 158 degrees 13 minutes West of True North;

South African line of masts: 72 degrees 5 minutes East of True North.

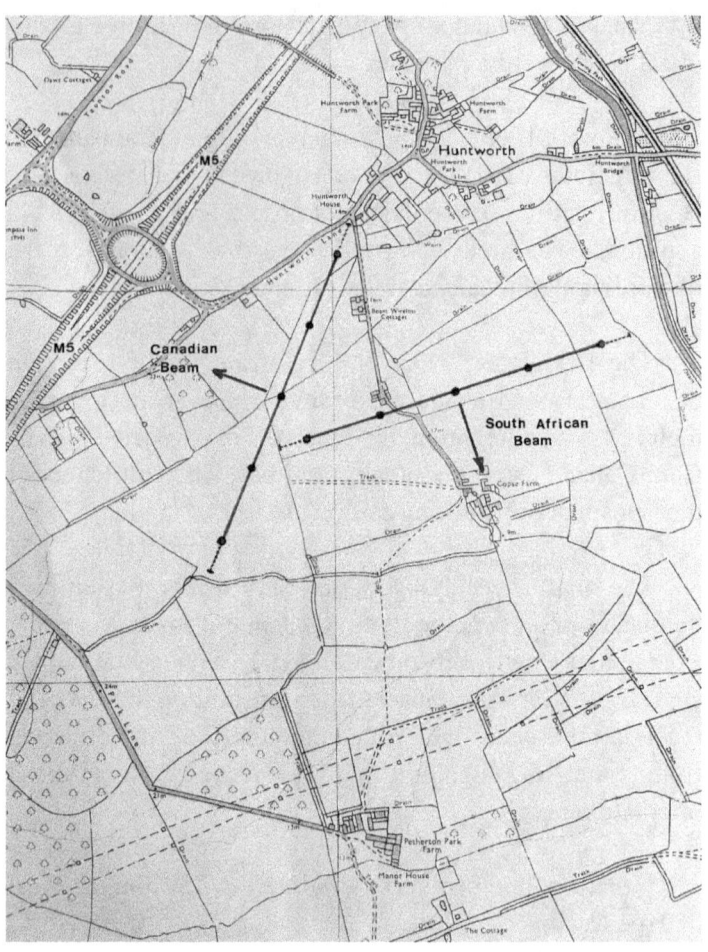

The location of the two Bridgwater Beam aerials,
plotted on a recent map of the area.

The effect of the reflector wires is not only to screen the aerial from signals coming from behind, but also to reflect back to the aerial energy received from the front of the system. Providing that the incoming signal from the front of the system is of the wavelength for which the aerials and reflectors are tuned and spaced, the energy received upon the latter is reflected back on to the aerial exactly in phase with that directly induced. This affords a very considerable increase in the energy received as compared with a broadcast or non-directional aerial.

By means of transformers and the special design of the feeder system, the energy of all the aerial wires is added and the total energy conveyed to the receivers, via the main feeder tubes, which terminate at what is known as a feeder unit of the receiver.

There are two Marconi short-wave beam receivers at Bridgwater, which are used for reception from two directions, namely from Canada and from South Africa.

The Feeder system at the Bridgwater station, showing the balance weight system which keeps the aerial and reflector wires taut and steady under wind pressure, and the coupling boxes by means of which the aerial system is coupled to the feeder system.

*Receiving equipment at the Bridgwater Station.
There are two distinct receivers shown here. The one on
the left is for South African signals, while the right-hand
one is for Canadian signals.*

In order to avoid interaction between the circuits, as they are worked on a common battery system, the frequency of the bands of the first filter amplifiers and two second filter amplifiers are suitably spaced. Otherwise the receivers are identical, and a description of one them is given below. The complete receiver comprises nine units, each contained in a copper screened box to prevent interaction between them; the general description of these units, the functions of which will be described below, being as follows:

1. Feeder terminal unit.

2. Modulator.

3. Incoming short-wave circuits and first heterodyne.

4. First filter amplifier.

5. First rectifier, second heterodyne and first listening circuit.

6. Second filter amplifier.

7. Additional second filter amplifier.

8. Main rectifier and second listening circuit.

9. Recorder and limiting circuits.

Receiver unit as used at the Bridgwater receiving station, showing the various modules used in the design of the equipment.

These units are mounted in a vertical iron rack or frame, and any unit can easily be changed if the need should arise. This method of mounting the units has the additional advantage of enabling the receivers to be adapted for any special purpose by the simple expedient of replacing one or more units without disturbing the general layout or the other parts of the set. The connections between the units are made at the back, and tuning controls are fitted on the front of the units, as well as switches and listening points which provide an audible check on the incoming signals. A control panel mounted on the left of each receiver enables the various voltages for the valves in circuit to be tested. A valve socket is provided to check the characteristics of any valve with the aid of the measuring instruments on the control board. The connections from filament and anode batteries as well as from the feeder are brought in at the back of each receiver.

The receivers are so designed that their speed of working is only limited by the mechanical means of transmission and reception available.

The feeder system at the Bridgwater Station.
The tubes on the left carried signals from Canada,
and those on the right from South Africa.

The feeder terminal unit, at which the aerial system is joined to the receiver, has two low-loss intermediate circuits, with a possibility of very weak coupling between them. This affords a means of obtaining additional selectivity and freedom from atmospheric and other interference. It is customary to work with very loose couplings in the unit, as, generally speaking, the signals are very powerful and the loose coupling facilitates the delivery of clear-cut signals to the lines connecting with the C.R.O. in London.

The second intermediate circuit is coupled to the input circuit of the receiver through a variable coupling. This input circuit is tuned to the frequency of the incoming wave, and is connected to the grids of two modulating valves, working in push-pull, i.e. one grid is positive when the other is negative."

However, not everyone welcomed the commencement of the new service. Radio amateurs in particular were quick to complain about the interference allegedly caused by these new stations to their radio installations, as a *'Popular Wireless'* article from December 1926 reveals:

"It is interesting to note that since the opening of the Beam Wireless stations at Bodmin and Bridgwater, many complaints of interference have been made by radio amateurs in the West Country. In many cases they report that their contact with the various broadcasting stations has practically been cut off and they suggest this is due to the beam stations when in operation.

One reader wants to know if the Postmaster General will explain how it is that on November 14th, when the beam station closed down in the evening, his reception in Cornwall was perfect, adding that last winter no interference at all was noticed and was considered that the beam station at Bodmin is causing all the trouble? Another listener states that the engineers of the B.B.C. who made investigations in Cornwall recently were satisfied that there is some interference which justifies the complaints of Cornish listeners.

The Canadian circuit at the Central Radio Office, London, 1926.

The same writer has made many interesting tests, the results of which, he claims, prove conclusively that the interference is caused by the Bodmin station. It seems here that the Postmaster General's attitude in regard to this interference is not quite so strong as it should be, and we suggest that further investigation of affairs in Cornwall should immediately be made, as it would be distinctly unfair if thousands of listeners in that part of the country were to have their wireless reception ruined owing to the operation of a beam station. We ourselves will welcome detailed reports from amateurs in the West Country who can show evidence that the interference is due to a beam station."

Bodmin, being the associated transmitting station, would naturally have the potential to cause interference, but it would have been unlikely that the Bridgwater receiving station would have had such issues.

It was still very exciting for the press to report on the new system and a call to Ottawa on 9th January 1927 was considered newsworthy enough to make the national newspapers:

"Shortly after noon on Sunday, says Reuters, direct wireless communication was established between Ottawa and Bridgwater, England. The head of the local tramways spoke with the Marconi engineer, communication being relayed via trunk telephone to Drummondsville, Quebec, and thence by Marconi Beam Wireless to England."

Another well-publicised event took place later that month when the previously-mentioned columnist *'Jack Broadcaster'* reported that:

"From Canadian papers that reach me, they are undoubtedly expecting to hear the King's voice when they celebrate their Diamond Jubilee on July 1st. A Canadian committee is planning the most extensive chain station broadcast ever attempted. They hope to have the carillon bells of the Peace Tower in Ottawa and a message of King

George, which will be made in reply to the playing of the national anthem by the bells, heard all around the world.

The Marconi-Mathieu Multiplex Telephone Receiving Apparatus at the Bridgwater Receiving Station. Mr. Matheiu, who collaborated with Senatore Marconi in the development of this apparatus is speaking on the radio-telephone.

The first broadcast is expected to take place at about 4 p.m. on July 1st. The programme will be relayed from Ottawa via telephone lines to Drummondville, Quebec. Here it will be picked up by a Marconi beam station and sent to England. Telephone lines will carry the signals to London, from where they will be broadcast throughout Europe. The message from the King will follow, by the reverse route."

'Wireless World' reported that the tests with South Africa were progressing well in their issue of 8th June 1927:

"The Marconi Company has completed the preliminary tests with the beam stations at Bodmin and Bridgwater which are being erected for the G.P.O. for communication with South Africa. During these preliminary tests speeds of between 200 and 250 words per minute have been maintained with South Africa over long periods daily. Further tests by the company under ordinary traffic conditions are now being made and if these prove satisfactory the stations will be handed over to the General Post Office for the official seven days' trial."

The South African service was formally opened on 5th July 1927 and was widely reported:

"The Anglo-South African beam service, which will be opened for public traffic at one a.m. today, places the Central Telegraph Office, London, in direct communication with the office of the Wireless Telegraph Company of South Africa in Cape Town.

A spectacular view of the Canadian and South African beam aerials and masts at the Bridgwater station.

Telegrams addressed to Cape Town are delivered to the public by the company's messengers. From Cape Town connection is given over the inland telegraph system to all destinations in the Union of South Africa, Rhodesia. South-West Africa, Nyasaland, Kenya, Uganda, Tanganyika Territory, Portuguese East Africa, Territory of Ruanda-Urundi, and Belgian Congo.

The rates of charge by this service to the Union of South Africa, Rhodesia, South-West Africa and Nyasaland are lower than those hitherto offered by any other route. No urgent or weekend service is being offered by the South African beam service at the outset, but the beam rates for daily letter telegrams to destinations are lower than the corresponding weekend cable rates by 1d a word. The beam rates are in all cases appreciably less than the corresponding cable rates. To secure the advantage of these reductions, telegrams must be marked 'Via Empiradio.'

Full particulars of this service can be obtained from the Postal Telegraph Office.

The King's Message

The radiogram from the Governor-General to the King, sent at the formal opening yesterday, read:

> I desire, in conjunction with my Ministers humbly to address your Majesty, on the occasion of the opening of the first Beam Wireless station in South Africa, a renewed expression of loyalty and devotion on behalf of your subjects in this Dominion. It is cause for rejoicing that in this improved system of communication we have proved that modern science, in its mastery over time and space will tend to draw closer the widely scattered nations of the British Empire, who look to your Majesty as their constitutional Sovereign.
>
> Athlone.

The King duly replied:

> I have received with sincere pleasure your message conveying an expression of loyalty and devotion on behalf of my people in South Africa on the occasion of the opening of the first Beam Wireless station there. Every invention of this kind, which is designed

to overcome distance is, I am convinced, of the greatest value in promoting mutual understanding and friendship between the nations of my Empire.

George. R.I.

The transmitting station in Great Britain for the South African service is at Bodmin, Cornwall and the receiving station is at Bridgwater. Both are controlled from the Central Radio Office at the G.P.O., London. The new stations are the first to put into operation the principle using two different wavelengths for night and day transmission."

These tests seemed to go extremely well, and the formal opening of the commercial service to South Africa was reported in another *'Wireless World'* article dated 7th July 1927:

"The recent successful tests with the South African beam service showed that, using two wavelengths, the two stations are capable of carrying on a high speed duplex service between London and Cape Town for the greater portion of the day and night, although the contract called for only eleven hours per day. It is estimated by the Marconi Company that the stations are able to handle 160,000 words per day in each direction. The transmitting and receiving stations in this country are situated at Bodmin and Bridgwater respectively.

The South African transmitter is at Klipheuval, thirty miles north-east of Cape Town, while the receiver is at Milnerton, five miles north of the city. The exact wavelengths of the English station are 16.146 metres (day service) and

34.013 metres (night service). The African wavelengths are 16.077 and 33.708 metres for day and night respectively."

That same month, members of the Taunton and District Radio Society were granted a visit to the Bridgwater Receiving Station, details of which were duly recorded:

"Wonders of the Beam Wireless System were elucidated to members of the Taunton and District Radio Society when they visited the North Petherton Beam Station on Wednesday evening. Many tips were gleaned by the amateurs, but there is no need to be afraid of even the most ardent wireless fans endeavouring to install such a system in their own homes. For instance, who could imagine a wireless mast 270 feet high towering above town houses, or the largest room in the house being utilised for rows upon rows of uniform accumulators? Yet these were but two of the numerous sights with which the visitors were thrilled.

The North Petherton Station receives messages in Morse from Canada and South Africa, which are relayed to London, where the messages are translated by the usual Creed system.

The Beam System is stated to be developing satisfactorily, with increased business resulting from the success achieved. There is no suggestion that the cable is being superseded, in fact, the opinion is expressed that there is ample room for both systems. Modern life is producing a greater desire for quick communication, and, apart from an increasing number of messages at full rates, there is a growing inclination to take advantage of the daily letter telegram and the weekend telegraphed letter.

The Beam System with its lower rates is expected to play an important part in the development of the cheaper forms of communication, and experience already shows that it is attracting a considerable part of this business. But in regard to urgent traffic which is estimated to be three times greater now than it was before the war, the Beam System does not appear to have seriously affected the cable services.

Communication with Canada has been in operation at North Petherton for some six months but the service to South Africa was only opened on Tuesday. For each country there are five of the huge steel-constructed masts and they run parallel with those at the transmitting stations. The party from Taunton spent over an hour at the receiving station where Mr. H. S. Reece and Mr. McDonald, two of the officials, lucidly explained the apparently complicated system."

The local Taunton press took full advantage of the publicity surrounding the new station, and they fully hoped that the location would put Huntworth on the world stage in an article from July 1927:

"Huntworth, formerly an obscure hamlet of North Petherton, is now becoming a well-known place, for in addition to the beet sugar factory proposed to be erected shortly, it is also the receiving station of the Beam Wireless for Canada and South Africa, both of which stations are now in full operation. Huntworth now bids fair to become not only of local celebrity but also world-wide fame."

It seems that not only telephone calls were being handled over the network; it appears that on one occasion in 1928, music was transmitted from Montreal using the same equipment and aerials through which two simultaneous Morse telegraph messages were being sent. Apparently the music was received at 'full strength' and there was no hint of Morse interference, and the equipment used 'considerably diminished' the effects of fading.

This was reported in the *'Western Daily Press'* on 22nd June 1928, under the heading of 'Music and Morse by Beam':

"Bridgwater dances to music from Montreal. Remarkable possibilities for cheap Empire-wide wireless telephone service are opened up by the success of experiments which are being made at the Marconi Beam Station at Bridgwater. These experiments, it is stated, have given striking proof of Senatore Marconi's claim that wherever short wave Beam telegraph service is conducted a telephone service may also be established, while the new 'Multiplex' apparatus that has now been developed enables one set of apparatus and aerials to conduct simultaneous telephone and telegraph services.

During the tests at Bridgwater, a party of Beam experts listened to dance music which was being received from Montreal at the same time and with the same apparatus and aerials as two Morse telegraph services from Montreal. The music was received at full strength, was of excellent quality, there was no hint of interference, and it was impossible to detect that the lilt of the dance band was being transmitted from Canada on the same radio circuit as a high speed 'dot and dash' service.

Mr G. A Mathieu, the research engineer, who has been collaborating with Senatore Marconi in the development of the Marconi-Mathieu Multiplex system, said to a reporter yesterday:

"The principle of the Multiplex system is this; if you listen to a concert from a broadcasting station you receive simultaneously a different note from every instrument that is before the microphone. The trombone gives a very low note, a piano perhaps a medium note, and a girl singing gives a high-pitched note. If you can now imagine an instrument which can select each of these different notes and separate them you can understand how the Multiplex works."

It is due to the anti-fading properties of the new apparatus that we can get an absolutely constant volume of speech, music, in spite of the fact that you are naturally dividing the power of your transmitter, among the different channels of communication.

At present the Multiplex equipment at the Canadian Marconi Company's Beam station near Montreal and the new receiver at Bridgwater are the only instruments of this type available. The construction duplicate apparatus is, however, being hurried forward by the Marconi Company's works at Chelmsford, and it is hoped that within three months' time two-way Multiplex working will be achieved between England and Canada."

Ownership and leasing of the station had always been a political bone of contention, with the possibility of leasing the system to a private company being discussed. The same newspaper reported this on 27th November 1928:

"A Pig in a Poke," Says Commander Kenworthy. The Prime Minister informed Mr Wedgwood Benn in the Commons that the consent of the House would not be sought before the Postmaster-General leased the Beam Wireless system. The lease would be granted under the general powers conferred upon the Postmaster-General by the Post Office Act.

There was no necessity to submit for the approval of Parliament the details of the contract with the new communications company in reference to the sale of the cables. Then we are asked to buy a pig in a poke," remarked Commander Kenworthy.

"The purchase of the pig is not yet completed," said Mr Baldwin, amid laughter.

Although nothing definite has yet been decided, it is quite possible that in the near future a commercial beam telephone service may be inaugurated between Great Britain and Canada. Since the early spring the Marconi Company have been carrying out experiments between this country and Canada, and these have proved so successful in every way that stage of perfection is said to have been reached.

A Post Office official informed a correspondent that, as far as he knew, the Post Office authorities had not yet been approached by the Marconi Company. He concluded that the Company would be expected to take the first steps.

All that has happened so far is that for about twelve months the Marconi Company have been granted the use of the Bridgwater and Bodmin stations for the purpose of conducting experiments. The Bridgwater Station is used for sending messages, and that at Bodmin for receiving[1]. At present, the service between England and Canada is the only one in operation. Even if someone in Berlin or Paris wishes to communicate with Canada he first has to connect with London."

[1] *The author of this article was obviously mistaken – of course Bridgwater was the receiving station and Bodmin the associated transmitting station.*

However, all was not 'plain sailing' as the cost of the service was discussed in Parliament on 19th December 1928, in an exchange between Sir W. Mitchell-Thompson (Postmaster-General) and various Members of Parliament:

Mr. Wellock asked the Postmaster-General the total capital expenditure on the Beam Wireless system in the possession and control of his Department?

Sir W. Mitchell-Thompson: The total capital cost to the Post Office of the four Imperial Beam Wireless services was approximately £242,200. I should perhaps add that, as the hon. Member is doubtless aware, there is an annual royalty payable to the Marconi Company so long as any of their patents are employed in the stations.

Mr. Wellock: Is that the total cost of the beam service?

Sir W. Mitchell-Thompson The total capital cost.

Mr. Malone: Is it a fact that some of these stations are already let to the Marconi Company for wireless telephony experiments, and, if so, is any rental being charged?

Sir W. Mitchell-Thompson We are allowing the Marconi Company to conduct certain experiments at the Bridgwater Station.

Mr. Hardie: And the turnover in the first year of the experiment is greater than all the capital that is going to be sunk in it."

Radio enthusiasts the world over were keen to be able to hear these new stations, and one listener (one C. Andrews of Middlesex) wrote to *'Modern Wireless'* in February 1929 with his findings:

"Sir,

In your December issue of the *'Modern Wireless'*, W. L. S. wishes to know if any reader has heard any telephony below 16 metres. Well, I should like to mention that some few weeks ago I heard an American accent calling Hallo, Bridgwater" several times, following up with American news and gramophone records; and also I have heard, well below that - on, I should think, about 10 or 12 metres - two Morse stations. (Being unable to read Morse I cannot say who they are, but I should very much like to know.) I should also like to say I can get almost any station with perfect ease on the short waves, with very little reaction."

Many short-wave listeners and radio amateurs of the time were undertaking a lot of experimental work on the short-wave bands, and Marconi himself paid tribute to their work which paved the way to further understanding of propagation and radio characteristics.

March 1st 1929 saw another leap forward in technology when the Bridgwater station was involved in a combined telephony and telegraphy test, using the new 'Multiplex' system developed by the Marconi Company. The *'Marconi Review'* Company publication, gave full details in its March 1929 edition:

Sir Robert Donald (1860-1933).

"Sir Robert Donald, G.B.E., LL.D., who was Chairman of the British Imperial Wireless Telegraph Committee, wrote an important special article in the *'Daily Mail'* of March 4th, describing the duplex telephone test carried out between Canada and England on March 1st. This demonstration took place at the Beam Receiving station at Bridgwater in the presence of Sir Basil Blackett, K.C.B., K.C.S.I., Chairman of the newly organised Imperial Communications Company, Mr. David Sarnoff, Vice-President of the Radio Corporation of America, and the Rt. Hon. F. G. Kellaway, P.C., Managing Director of the Marconi Company.

The telephone conversation, which took place with the Vice-President of the Canadian Bell Telephone Company and the Vice-Chairman of the Canadian Marconi Company, who were in Montreal, was perfect, and Sir Robert Donald described the occasion as marking a revolution in wireless telephony - another triumph for the inventive geniuses of the Marconi Company.

Sir Robert, continuing his description, said: We were witnessing the successful demonstration of transmitting speech over a distance of more than 3,000 miles on the same wavelength simultaneously with two wireless telegraph messages. While we were using the desk telephone in the office of the station, the tape was running off written telegraphed messages in the operating room at the rate of hundreds of words a minute. These were being relayed to London without interruption.

The invention is a unique combination of efficiency and economy. The system may be incorporated in all 'Beam' stations on the Imperial Chain, avoiding the necessity for erecting new stations for telephonic communication.

The secret of the system is in reception. Up to now it has not been possible to use the same station for wireless telegraphy and telephony.

When messages are transmitted from Montreal simultaneously on the same wavelength they are sorted out on reception. While the Morse code messages continue without further assistance, the spoken message is regulated by other inventions, in particular by a gain controller and an 'echo suppressor'.

The advantages of the Marconi-Multiplex System, as it is called, are many," Sir Robert continued. "It is more efficient and incomparably cheaper. It does not involve the reconstruction of stations. It enables a short wave to be used. There is great economy in power, and there is almost complete secrecy. No one could have listened to our conversations except through a similar installation. Part of a message could be heard, but the whole conversation could not be picked up.

What is being accomplished on the Canadian circuit can be carried out between other parts of the Empire. Already an apparatus is being made for the South African Beam station. The universal adoption of the system would enormously cheapen telephonic communication throughout the Empire.

These new developments, which have now been brought to a commercial stage, have taken eighteen months to mature. They will make the British Empire more independent in its means of communication, and once all the home and overseas Beam stations are equipped with new reception apparatus conversations between England and the Dominions, will become almost as usual as they are today between London and the Continent.

This opinion of the Marconi-Multiplex System from an independent observer such as Sir Robert Donald, whose experience as Chairman of the Imperial Wireless Telegraphy Committee makes his views of unusual weight, will be of the greatest value to all who are interested in the establishment of wireless telephone services over long distances."

4th September 1929 saw the Beam Wireless Service come under the sole control of Imperial & International Communications Ltd., having previously been operated under the auspices of the Imperial Cables and the Post Office Central Telegraph Office.

The Beam Wireless sites owned by the Post Office were leased to the new company under a detailed lease agreement on that date between His Majesty's Postmaster General and Imperial and International Communications Limited, regarding the lease of the stations at Bodmin, Bridgwater, Skegness and Grimsby. A term of 25 years from 1st April 1928 (determinable as within) was the basis of the original agreement, with a rental of £250,000 per annum and further rent, as within, in addition. Inspection of the lease agreement regarding the transfer of staff and the appropriation of circuits to the stations shows that all due diligence had taken place on both sides, and very little was left to chance, although the agreement was to prove controversial some years later.

Extracts from the agreement show the various tenancies and covenants which affected the use of the land covered by the station, and make for interesting reading (and yes, the agreement does indeed locate North Petherton in the County of Cornwall!):

Description of Premises

All those pieces or parcels of land situate in the Parish of North Petherton, Bridgwater, in the County of Cornwall containing in the whole by admeasurement 240.114 statute acres or thereabouts together with the Wireless Station buildings, erections and masts now in upon or about the said pieces or parcels of land collectively form the premises known as the Bridgwater Beam Wireless Station and are now partly in the occupation of G. K. Norman, C. W.

Norman, the Rev. P. T. Pryce-Mitchel and Messrs. John and Alec Venner as tenants thereof.

Cover page of the 1929 Lease Agreement.

Agreements, Stipulations and Conditions

The aforesaid agreements, stipulations and conditions are contained in a deed of mutual covenants dated 20th day of December 1920 and made between Walter Edward Ludlow of the first part, Ernest Hine of the second part, Albert Arthur Broughton of the third part, and Thomas Bond Chapman of the fourth part, as varied by an Indenture of Conveyance dated the 29th September 1921 and made between the said Ernest Hine of the one part and Christopher William Norman of the other part.

The Lessee has been supplied with a full copy of these deeds and the accompanying plans and has full knowledge of their contents.

Restrictive Covenants

These restrictions and conditions are contained in a Conveyance dated the 5th day of June 1925 and made between the Right Honourable Claud Berkeley Fourth Viscount Portman of the first part, Edwin Wilfrid Stanyforth and the Honourable Egremont John Mills of the second part, and His Majesty's Postmaster General of the third part.

1. The Postmaster General, his successors or assigns shall not carry on or permit to be carried on the said premises or any part thereof any noxious, noisy or offensive trade or business but the user of the same for the purpose of or in connection with a Wireless Station shall not be a breach of this stipulation.

2. The Postmaster General, his successors and assigns shall not use any building now erected or hereafter to be erected on the said premises or any part thereof as a beerhouse or public house or for the sale of intoxicating liquors.

3. The Postmaster General, his successors and assigns shall not use or permit the said premises or any part thereof to be used as tea gardens or a place of public amusement nor allow any caravan house on wheels, show booth, swings or roundabouts to be placed thereon.

4. In the use of the said premises as a Wireless Station, the Postmaster General, his successors and assigns shall erect all necessary buildings, standards, posts, stays, aerials, machinery and works in such a position or manner and carry on the work of a Wireless Station in such manner as to cause the least possible annoyance, disturbance, damage or depreciation to the Vendor's adjoining property on the north-east side thereof known as Huntworth House and grounds or to the tenants or occupiers thereof or the amenities thereof.

Tenancies

Let to:	Cultivation.	Total Area in acres.	Terms of Occupation.
Mr. G. K. Norman	Pasture	13.982	Annual Michaelmas Tenancy at £33 5s 0d p.a. exclusive of rates.
Rev. Pryce Mitchell	Pasture	6.167	Let on annual Michaelmas Tenancy at £9 p.a. exclusive of rates.
Mr. C. W. Norman	Pasture	8.863	Let on annual Michaelmas Tenancy at £10 p.a. exclusive of rates.
Copse Farm, let to Messrs. Venner	Pasture, Farmhouse buildings, yards, and arable pasture	171.876	Let on annual Michaelmas Tenancy at £380 p.a. exclusive of rates. Tenants have right of way over Park Lane and roadway through adjoining land (not Post Office property).
Wireless Station Buildings, Drive, and Two Aerial Enclosures	Drive, Official Residences, Aerial Enclosures, and Station Building	240.114	

News of this change of organisation appeared in the *'Telegraph and Telephone Journal'* of September 1929, and reported in the somewhat evocative language of the time:

"For a time the Imperial Cables will be worked from Electra House, Moorgate, the headquarters of the Eastern Company, and the Beams from Radio House, Wilson Street, the main telegraph office of the Marconi Company. At a later date the services will meet again at the new headquarters of the Communications Company, which are now being built on the Embankment.

This journal, by reason of its habitual modesty and discretion, has done much less than justice to the Beam Services, which may be described without exaggeration as the most important development in the history of telegraphy during the present century. In the early days of Beam working the system was untried, and a large staff had to be specially trained to adapt their methods to the peculiarities of the new system. In these circumstances it was premature to boast. When the Beams had begun to establish themselves, the shadow of the Imperial Conference loomed on the horizon; and, although the daily press was full of rumours of what the Conference were likely to decide and the shares of the Eastern & Marconi Companies 'soared' in consequence, it behoved us to keep silence until the report of the Conference was published. A summary of that report will be found in our issue of August, 1928; and at last, after long and complicated negotiations, the recommendations of the Conference have fully taken effect.

Let us examine briefly in retrospect the Post Office working of the Beam service (considerations of space alone compel us to crowd out the less spectacular but hardly less significant history of the Imperial Cables). First of all from the taxpayers' point of view. The capital cost involved to the Exchequer in the four Beam Services in this country was about £242,000. In the first complete year since their inception, the working profit gained by the Post Office amounted to no less than £166,000; and the present profits are still higher. The financial exploitation of a new invention is strewn with pitfalls, as investors in any new companies have found to their cost; and the result in this case is a record of sound financial administration of which we may be reasonably proud.

Next from the point of view of the 'consumer'. Mainly as a result of the instance of the British Post Office, it was decided at the outset to fix the main rates on the Australian, Indian and South African services at two-thirds of the cable rates then in force; and on the Canadian service, where it was not found possible to reduce the rates generally, a special service of 'post letter telegrams' was introduced at the unprecedentedly low rate of half-a-crown (12½p) for a message of twenty words to any part of Canada. Immediately before the Beam Services to Australia, South Africa and India were opened, the cable rates to those countries were also reduced in order to meet the new competition.

These solid gains have been secured to the British public, not only by reason of the invention of the Beams, but also to no small extent by virtue of the fact that the Beam Services in this country were controlled by the Post Office.

As regards the confidence of the public, the traffic speaks for itself. If we contrast the first complete week of each service with a week in November last, chosen at random (the latest comparison that has been published), we find that on the Canadian Beam the traffic increased from 59,000 to 113,000 paid words; on the Australian Beam from 53,000 to 181,000; on the Indian Beam from 115,000 to 253,000; on the South African Beam from 88,000 to 200,000. The proportion of full-rate traffic to the whole is about 50%. It is clear that the business community is not only making use of the Beam Services but is ready to entrust them with its code traffic, for which speed and accuracy is essential.

As regards the staff, the operating and engineering staff directly concerned in the working of the services have been given an opportunity of transferring to the Communications Company. Some have decided to avail themselves of this opportunity; the majority, as far as known at present, have preferred to remain with the Post Office. But whether they go or stay, they will not lightly forget their experiences of the last two years. They worked at concert-pitch, in the full limelight of criticism; and they made a good show.

The services established a sound tradition; and it is to be hoped that even though the services are lost, the tradition will remain. On the remainder of the telegraph service there is not the same incentive of novelty and competition; but on the other hand a monopoly imposes an obligation, and the public have no less right to a first-rate service because they have no opportunities of transferring their custom elsewhere.

We trust, therefore, that the telegraph service will profit by the association of the Beams with the Post Office, not only by the annual rent of £250,000, but in other less material ways. There is some hope at any rate that the publicity organisation (commonly known as 'canvassing'), to which the Imperial Cable and Wireless services owed so much, will be retained; and the operators who were trained to give a high quality to Melbourne, Bombay, Cape Town, and Montreal will not be content with an inferior service to Manchester or Paris.

We had built up high hopes of the Imperial services, and it is with real regret that we hand them over; but we wish every success to the Communications Company in this extension of the activities. We are especially sorry that the team of operators at the C.T.O. will have to be disbanded; both to those who leave us and to those who remain we extend our sincere congratulations on the past and our cordial wishes for the future."

October 1929 saw an excellent review of the Beam Service network published in the *'Marconi Review'*, which devoted a section to the Bridgwater station:

"The two receivers for the Canadian and South African services occupy the middle of the picture, and in front of them are the undulators of the monitoring circuits with line apparatus to enable the operators on watch to communicate with the Central Telegraph Office or the Transmitting Station at will. The Multiplex apparatus is in the rack on the left. This is not yet part of the standard equipment, but is at present receiving all the beam telegraph traffic from Canada in place of the usual receiver, and its second channel is employed for beam telephony tests as and when required. The third channel is used for telegraph traffic when necessary."

The process of sending multiple messages and telephone calls on the same circuit was a revelation, and was reported across the UK at the end of the year in glowing terms:

"Further experiments are to be made by the Marconi Company with a view to demonstrating the possibility of sending two or three telegraphic messages and a wireless telephone message at the same time over the same circuit on the one set of apparatus by short wave Beam Wireless.

Since experiments were inaugurated last June between Montreal and Bridgwater, Somerset, they have been confined to the receipt of messages from Canada to England.

Apparatus has now been sent to Canada to enable transmission and reception to be carried on from both sides of the Atlantic. "One-way tests have been very satisfactory." an official of the Marconi Company said yesterday.

This new system would enable all Beam Stations to deal with three times the amount of their present traffic.

This new apparatus has also overcome the difficulty of fading, a trouble associated with the normal type of short wave working."

***Receiving room at the Bridgwater station, with a
Marconi-Mathieu Multiplex Receiver and two Beam
Receivers.***

Returning to the politics going on behind the scenes,
it appeared that the Post Office were not best pleased with
the new arrangement, and co-operation with the Imperial
and International Communications Company can best be
described as 'grudging acceptance'. The Post Office still
owned the landlines to and from the stations, and this
was apparently used as a 'bargaining tool' with the new
company. In fact, it was reported in the House of Commons
on 26th March 1930 during a speech by Mr. Leo Amery,
M.P. for Birmingham Sparkbrook, that:

"As far back as February, 1929, when the arrangements
for handing over were complete, the Post Office assured the
company that it would not place any orders for apparatus
for Dominion services for at least six months without

consulting the company, and it undertook to co-operate with the company with a view to the development of the most useful apparatus. On that, one might have supposed that negotiations and discussions would begin, but nothing happened, although, as early as July of last year, the Post Office, without informing the company - who, after all, ought to be its partner in this Imperial business, and not an enemy to be kept at arm's length - wrote to the Indian Radio-Telegraph Department saying that it was very desirable to have a system of communications independent of the Communications Company.

Obviously, their whole attitude was to try to get away from the company, and the offer of the company to make an immediate opening of services with Canada and other Dominions, instead of being met by a request to come round and discuss and consider the matter at once, received no answer whatever except a formal acknowledgment. As far as the Post Office was concerned, we do not know when any answer would have been received by the company. It was only when the company appealed direct to the Chancellor of the Exchequer, protesting against the attitude of the Post Office that something at last happened. These protests were made at the end of September, and again on the 1st October, when the company felt bound to inform the Chancellor of the Exchequer that the company's telephonic communication at its station at Bridgwater had been cut by the Post Office, and had been taken up by the Post Office without even the courtesy of a warning or intimation, thus making it impossible for the company even to carry on experimental trials in order to show foreign visitors what it could do.

The whole attitude was one of deliberately preventing the company not only from carrying on a commercial service, but even from carrying on experiments in this country."

Mr. Amery went on to say:

"The Postmaster-General repeated today what he said on 26th February, that concentration at Rugby admits of economies in many directions, and particularly in the land line connections to the London Trunk Exchange. Are there no trunk lines already in existence within a very close distance of the company's station? Is it not a fact that there was actual communication between Bridgwater and the main trunk line, which the Post Office deliberately severed as part of their policy of discouraging the company?"

It was subsequently confirmed in March 1930 that the Government were considering concentrating all the Wireless Beam Stations from one transmitting location at Rugby, with a receiving station at Baldock, which would reduce the landline distance to and from London.

'Wireless World' reported this consideration:

In the House of Commons on Wednesday last, **Mr. Bowen** asked the Postmaster-General whether any decision had been reached as to the control of Imperial wireless telephony and whether it was intended to use the Beam Stations.

Mr. Lees-Smith (Postmaster-General): 'Yes, Sir, the Government have reached a decision, and, with the permission of the House, I will state briefly the main reasons for it. Under the late Government the Beam Wireless system for overseas telegraphy was leased to the Imperial and International Communications Company under conditions and circumstances which are well known. The late Government, however, in conformity with the recommendation of the Imperial Wireless and Cable Conference, reserved to the Post Office the control of overseas telephony and deliberately refrained from committing themselves on the question whether they would or would not use the company's stations for this purpose.

In August last I received a letter from the Communications Company urging that the Government should now decide to work overseas telephony through the company's stations, beginning with four services to Canada, Australia, South Africa, and India. This was one alternative. The other was to concentrate all their wireless telephone services at the Government station at Rugby, which has for three years worked the service to the United States on a commercial basis. In deciding between these two alternatives there were two main issues: first, which of the two systems would provide the most efficient service; and, secondly, which would be the more economical.

As the first question involved highly technical considerations, the Government decided to consult two independent experts of acknowledged repute, who have no connection with the Post Office. Professor G. W. Howe, Professor of Electrical Engineering at the University of Glasgow, and Dr. F. E. Smith, Secretary of the Royal Society and of the Department of Scientific Research. They reported that apart from future developments both systems are probably equally capable of providing satisfactory telephonic communication between two points for a given number of hours a day, and that, as regards future development, the Rugby system was the more elastic and therefore in this respect offered decided advantages.

The second main issue is the financial comparison between the two systems. Concentration at Rugby admits of economies in many directions, and, in particular, in the landline connections to the London trunk exchange. A wireless service requires costly landline connections between the London trunk exchange and the wireless stations. By grouping of services at one centre, such as Rugby, a smaller number of lines will suffice and the distance of Rugby and Baldock from London is much less than the distance of the beam stations at Bodmin, Bridgwater, Grimsby, and Skegness: The result is that to work the four services to India and the Dominions through the beam stations would need 4,190 miles of high-grade telephone circuit, while to work them through Rugby and Baldock, only 786 miles would be required.

The minimum rental asked by the company for the use of the beam telegraph stations for the telephone services in question is (excluding a cheaper scheme which is open to objection on other grounds) £40,000 to £45,000 per annum, according to the type of equipment employed, plus a royalty of 10 per cent on the gross receipts in excess of a certain figure.

A detailed estimate of the cost of working the same services from Rugby shows a saving on the above figures of £17,000 per annum and £22,000 per annum respectively, which would be increased when the royalty commenced to operate.

The Government has had to weigh the pros and cons of a number of other questions which cannot be compressed into a Parliamentary answer. As a result of their consideration of all the issues they have decided upon a policy of conducting overseas wireless telephony by concentration at the Post Office station at Hillmorton, hear Rugby, with its receiving station at Baldock."

Further details of the changes (some shown earlier in this chapter but repeated for the sake of clarity) were promulgated in the national press:

OVERSEAS RADIO TELEPHONY. DECISION OF THE GOVERNMENT. CONCENTRATION AT RUGBY & BALDOCK. FUTURE OF 'BEAMS' IN THE WEST

"The Government has decided upon a policy of conducting overseas wireless telephony by concentration at the Post Office stations Rugby and Baldock.

This important announcement, made by the Postmaster-General (Mr. H. B. Lees-Smith) in the House of Commons yesterday, is of particular interest to the West Country in view of the fact that the counties of Cornwall and Somerset have figured prominently in the greatest wireless development of recent years - the beam system.

Two groups of stations on the beam system were erected for the Post Office by the Marconi Company under contract.

The first, for communication with Canada and South Africa, has its transmitting unit at Bodmin and its receiving unit near Bridgwater. The second, for communication with Australia and India, has its transmitting and receiving units at Grimsby and Skegness respectively. In future, these beam stations will not be used by the Post Office for radio telephony to the Dominions.

In a subsequent statement, issued last night, the Postmaster-General said the question at issue was whether the services should be worked through the beam telegraph stations or concentrated at the Government transmitting station at Rugby, with its complementary receiving station at Baldock. The experimental tests with Australia, which the Post Office had carried on through Rugby, had more than fulfilled expectations.

Two independent experts were consulted, and in the view of the experts' report, the Government were justified in concluding that without disparaging the possibilities of the Marconi system, at least as efficient services could be provided from Rugby, with certain economic advantages in prospect.

"There is no intention," he added, "of duplicating the Marconi stations, and the total cost of the plant required for each service, including both transmission and reception, will not be £112,000 or more, as has been stated, but slightly under £30,000.

"It is not true, as has been alleged, that the use of Rugby would almost certainly entail the erection of corresponding stations in each Dominion."

The Postmaster-General's statement in the House of Commons was made in answer to a private notice question by Mr. Bowen, Labour member for Crewe. Mr. Lees-Smith reminded the House that under the late Government, the Beam Wireless system of overseas telegraphy was leased to the Imperial and International Communications Company under conditions and circumstances which were well known. The late Government, however, in conformity with the recommendation of the Imperial Wireless and Cable Conference, reserved to the Post Office the control of the overseas telephony, and deliberately refrained from committing themselves on the question whether they would or would not use the company's stations for the purpose.

"In August last," he proceeded, "I received a letter from the Communications Company urging that the Government should now decide to work overseas telephony through the company's stations beginning with four services to Canada. Australia, South Africa and India. This was one alternative.

The other alternative was to concentrate all their wireless telephony services, at the Government station at Rugby, which has for three years worked the service to the United States on a commercial basis.

In deciding between these two alternatives there were two main issues; firstly, which of the two systems would provide the more efficient service and secondly, which would be the more economic.

As the first question involved highly technical considerations, the Government decided to consult two independent experts of acknowledged repute, who had no connection with the Post Office - Professor G. W. Howe, Professor of Electrical Engineering at the University of Glasgow, and Dr. F. E. Smith, F.R.S of the Department of Scientific Research. They reported that apart from future development, both systems are probably equally capable of providing satisfactory telephonic communication between two points for good number of hours a day, and that as regards future development, the Rugby system was the more elastic, and therefore in this respect offered decided advantages.

The second main issue is the financial comparison between the two systems. Concentration at Rugby admits of economies in many directions, and in particular the land line connections to the London trunk exchange."

A wireless service requires costly land line connections between the London trunk exchange and the wireless stations. By grouping of services at one centre, such as Rugby, a smaller number of lines will suffice, and the distance of Rugby and Baldock from London is much less than the distance of the beam stations Bodmin, Bridgwater, Grimsby, and Skegness.

The result is that to work the four services to India and the Dominions through the beam stations would need 4,190 miles of high-grade telephone circuit, while to work them through and Baldock only 786 miles would be required.

The minimum rental asked by the Company for the use of the beam telegraph stations for the telephone services in question is - excluding a cheaper scheme, which is open to objection on other grounds - £40,000 to £45,000 per annum, according to the type of equipment employed, plus a royalty of 10 per cent, on the gross receipts in excess of a certain figure.

A detailed estimate of the cost of working the same services from Rugby shows a saving on the above figure of £17,000 per annum and £22,000 per annum respectively, which would be increased when the royalty commenced to operate.

The Government has had to weigh the pros and cons of number of other questions, which cannot be compressed into a Parliamentary answer. As a result of their considerations all the issues, they have decided upon a policy of conducting overseas wireless telephony by concentration at the Post Office station at Rugby, with its receiving station at Baldock.

Mr. Baldwin said the statement was of very great interest and importance, as well as of great complexity. He was sure the Postmaster-General would welcome the opportunity of making a more detailed statement and he hoped the Government would give opportunity for discussion later on (Ministerial cries of "Why?").

Mr. A. M. Samuel (Con.), asked whether it was the intention of the Government to enter into partnership with any American foreign company with regard to these overseas communications, as, for example, with Egypt, and if the Postmaster-General proposed to ride roughshod over recommendation No. 8 of the Imperial Wireless Conference report.

Mr. Lees-Smith: We do not intend to override recommendation No. 8. It merely lays down that the Government should have the right if they wish to use the beam station, but it imposes no obligation upon them. We do not intend to enter into any partnership with any American or foreign companies, but at the other end of the services, which is in foreign countries, we are bound to make some arrangement.

Further answering Mr. Samuel, Mr. Lees-Smith said the Government could not say with what companies they would act in foreign countries. They had to consider which companies would give access to the greatest number of subscribers. The Government would do all it could to support British interests. Mr. Baldwin renewed his suggestion for a further discussion at a later date, and Mr. A. Henderson, Foreign Secretary, said he would convey the suggestion to the Prime Minister."

On Saturday 8th November 1930, a delegation from the Imperial Conference visited the Bridgwater station, as well as the accompanying transmitter station at Dorchester. A copy of the souvenir programme will be found in Appendix 2.

This high-profile visit was well reported locally:

"A number of delegates to the Imperial Conference, accompanied by Mr. J. H. Thomas, Secretary of State for the Dominions, visited the Imperial Beam Wireless Stations at Dorchester and Bridgwater yesterday and watched the reception and transmission of messages to all parts of the world.

Bridgwater and Dorchester are the hub of a network of short-wave Beam Wireless, which, coupled with cables, connects the utmost corners of the world with England. Arrangements were made for Mr. Thomas and Mr. Scullin. Prime Minister of Australia, and other members of the party to exchange greetings with friends in Montreal and Ottawa by means of wireless telephony.

Sir Basil Blackett, chairman of Imperial and International Communications Ltd., presided, at Weymouth, at a luncheon which followed the visit to the Dorchester station. He said the one subject of enduring interest which was seen to occupy each and all of the Imperial Conferences was the subject of unity within the Empire, and of this unity Imperial and International Communications was not merely a symbol but an agent. Their sound was gone forth into all lands - their words unto the ends of the earth. It was by their means that Dominion exchanged thought with Dominion, Colony with Colony, and each of them with the Mother Country.

The visitors had that day talked on the telephone to Canada and by telegraph to South Africa. The Dorchester station, which they had seen, and the Bridgwater station, which they would see that afternoon, were two of the first stations erected to work on short waves with Beam Aerials. Those stations had revolutionised wireless, for until they were erected, no commercial stations utilising short waves and directive wireless had as yet been used. The leadership then held - for those stations were built as long ago as 1926 - had since then been fully maintained.

"The stations have been kept up-to-date, and every useful invention or improvement which has been developed during the past years has been or is being installed. I give you a few random examples. Substantial improvements have been made in Beam Aerials, resulting in greater concentration of the rays in the vertical plane. Modulation has been provided to the transmitters, thereby reducing rapid fading. Adaptation has been provided so that telephony can be at once made use of where required.

Methods have been adopted whereby multiplex working can be effected, enabling two telegraph channels and one telephone channel to be served from the same transmitter simultaneously. The constancy of the carrier frequency - which was already very good – has been or is being substantially improved."

Then there was facsimile. Facsimile had just been installed on the South African circuit and the Canadian circuit, and would shortly be installed on the other Imperial circuits. This enabled a cheque or any document to be faithfully reproduced at the receiving end, no matter what the distance might be between the two stations.

A few weeks ago the Beam from Australia transmitted experimentally a photograph of Commander Kingsford-Smith on his arrival in Australia. It was published in England. A few days ago a letter from Mr. Bennett was sent in facsimile to Canada by the Canadian Beam. The Empire had been provided with the finest system of wireless communications in the world.

Imperial and International Communications carried in 1929 104,000,000 words between Empire countries, by cable and wireless, and 73,000,000 words between Empire and foreign countries, out of a total of 244,000,000 words. They felt a special responsibility to try to make their company an organ of economic co-operation in the Empire."

The visit was also reported in more detail locally:

"The remarkable strides made recently in wireless telephony were demonstrated to the Imperial Conference delegates at the Beam Wireless Station at Huntworth, North Petherton, on Saturday. In a large marquee erected adjoining the Beam Station, several of the visitors spoke to friends across the seas, and their replies, which were heard through a loud speaker, were as distinct as if the voices were actually in the tent instead of thousands of miles away.

Included in the party, which numbered over 100, were the Right Hon. J. H. Thomas, M.P., and Mrs. and Miss Thomas, the Marchese E. Marconi, G.C.V.O., LL.D., D.S.C., and the Marchesa Marconi, Sir Basil Blackett, K.C.B., K.C.S.I., and Lady Blackett, the Lord Mayor and Lady Mayoress of Bristol, and many other distinguished people.

Mr. Thomas, speaking to Mr. E. W. B. Beatty, President of the Canadian Pacific Railway, declared emphatically that the 1930 Imperial Conference was not going to break up without some practical results.

"That's good news" answered Mr. Beatty.

Sir Basil Blackett, Chairman of Imperial and International Communications Ltd., speaking at the luncheon which the party attended, remarked that the one subject of enduring interest, which was seen to occupy each and all of the conferences, was the subject of unity within the Empire."

As regards to the visit to Bridgwater station alone, a fascinating document recalls some of the events of the day:

"Hello, that you Beatty? How are you?"

"I am all right, how are you?"

"How am I? How the deuce do you think I am!"

Of the voices engaging in the above colloquy one was that of the Rt. Hon. J. H. Thomas M.P., Secretary of State for Dominion Affairs, who was speaking into a telephone in a tent at Huntworth, near Bridgwater, on Saturday afternoon. The other – the answering – voice, heard with great distinction in the same tent, was that of a railway agent speaking from his office in Montreal.

The questions preceded a long conversation between the two by means of wireless telephony at the Bridgwater Beam Wireless Station, a demonstration of the remarkable strides made in the development of wireless telephony being given in the presence of the distinguished gathering which included the Marchese Guglielmo Marconi G.C.V.C., LL.D., D.Sc., the 'wizard' of wireless, who had paid previous visits to Huntworth.

The occasion of the experiment was the visit to the Beam Station of Imperial Conference delegates at the invitation of Cable and Wireless Ltd., and Imperial and International Communications Ltd.

The party had left Bristol shortly before nine o'clock in the morning, proceeding by train to Dorchester, where they visited the Beam Wireless Station. Saloon road coaches then conveyed them to Weymouth for luncheon, and they afterwards came to Huntworth by road coaches, The Rt. Hon. J. H. Thomas, who was accompanied by Mrs. Thomas, arrived in a motor car, and alighted smoking the ubiquitous cigar.

The Beam Station was reached about 4.30 p.m. and immediately on arrival the visitors posed for press photographers and then without more ado proceeded to a marquee which had been erected close to the Beam Station. The marquee had been beautifully fitted up as a lounge by Messrs. W. and A. Chapman of Taunton, flags adding to the adornment.

Mr. Thomas (seated) speaks to Canada from Bridgwater, with various dignitaries in the background, including Guglielmo Marconi and Sir Basil Blackett.

None of the Colonial Prime Ministers attended, but many other Imperial Conference delegates were present, as well as a number of journalists including representatives of Canadian and South African papers, the party numbering over 100.

When the visitors had filed into the marquee, the demonstration of wireless telephony commenced without ceremony. Mr. J. H. Thomas, as the chief guest, seated himself at a table upon which rested a small telephone box and lifting the receiver he was within a few seconds in conversation with Mr. E. W. B. Beatty, president of the Canadian Pacific Railway, who was in his office in Montreal. The Dominion Secretary's talk with Mr. Beatty was punctuated with laughter from the assembled company.

The wizardry of wireless telephony was remarkably demonstrated. The answers of Mr. Beatty were heard through a loudspeaker, from which had been emanating music from Ottawa and Montreal, and Mr. Beatty's replies were as distinct as if the voice was actually in the tent instead of the words being uttered several thousand miles away.

After the opening salutations, Mr. Thomas informed Mr. Beatty that he had had half a dozen of the Canadians bothering him for about six weeks, with a few Australians, Irishmen and New Zealanders thrown in.

The loudspeaker emitted a laugh.

"Oh you may laugh" said Mr. Thomas.

"But I have heard nothing about anything except wheat for the last six weeks (laughter). That is true. What do you think about things in general?"

"If you can report progress that is alright", came the answering voice of Mr. Beatty.

"That shows," replied Mr. Thomas to the accompaniment of laughter, "that you are a long way from the scene of action".

1936 saw the 10th anniversary of the first Empire Beam Wireless Service, which was celebrated by a short celebration in the national press on October 25th of that year.

"An official of Cable and Wireless Limited, who now operate the Beam Service, spoke to a representative of *'The Observer'* of the early days of one of the most fascinating of modern inventions, and the experiments of the young Marconi, and of how, in 1901, the letter 'S' was first flashed across the Atlantic, unmistakably distinct.

Then in 1923 when long-wave wireless seemed to have reached its limits because the ether was so crowded, the Marchese Marconi made his second revolutionary discovery – the short-wave Beam System.

The Canadian service of 1926 was soon followed by services between South Africa, Australia, and India. The control of Imperial overseas telegraphy was vested in Cable and Wireless Limited, who have transmitted over 1¼ million words in a week on the Beam circuits alone.

The operator in London is in complete control of the transmitter and, as an instance of speed, a message has been dispatched to New York, and the reply received in London within twenty-four seconds for the two. The official requirement of the Canadian stations was that they should be capable of communications at a speed of 500 letters a minute each way during a daily average of eighteen hours; actually, during tests, 1,250 letters a minute were transmitted for hours on end.

Looking ahead, the development of facsimile transmission has been so rapid that it may not be long before telegrams will be projected through space and reproduced at the other end in the sender's own handwriting. Already, apart from the many photographs of momentous events, fashion plates, cheques, and architects' plans are being wirelessed across the world.

In one case, a chart, with essential information, was transmitted to a cable ship in Melbourne, which was thus enabled to proceed to the repair of a damaged cable in South Africa. For, apart from the Beam Service, the Company operates more than half of existing submarine cables, which, as with wireless, made a beginning across the Atlantic."

Another high-profile event showcasing the Canadian service took place on January 1st 1931, when it was reported that:

"One of the most interesting events on New Year's Day will be the opening of the Schoolboys' Exhibition at the Empire Hall. Olympia. London, by the Lord Mayor of London (Sir W. Phené Neal) and the Lady Mayoress. The Lord Mayor's speech during the opening ceremony will be transmitted by Beam Wireless to Ottawa, where it will be heard by the Prime Minister of Canada.

The Canadian Premier's reply will arrive at Olympia by Beam Wireless within a few seconds, and will be amplified and broadcast by loudspeakers and heard throughout the Exhibition, at which every one of the three floors of Olympia's great Empire Hall will he packed with attractions and exhibits before which every boy and girl - and no doubt parents also - will want to linger."

The Bridgwater station beam aerials were deemed so impressive that they appeared in an official Cable and Wireless brochure published in the 1930s, entitled 'A Short

History of British Overseas Communications', extracts from which are shown below:

The brochure was keen to extol the virtues of the Beam Wireless system, and gleefully noted that "In one week 1,255,201 words were transmitted and received over the Empire circuits alone, and this volume of traffic was handled with the greatest ease owing to the phenomenal speed of the beam. Over 300 words a minute have been attained in commercial operation, this being easily record for any system of long-distance telegraph communication".

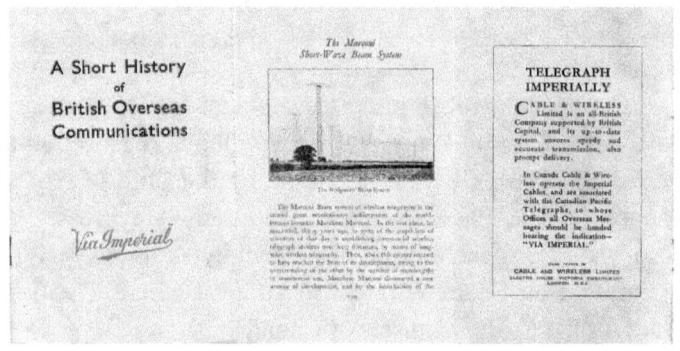

Selected pages from the aforementioned Cable and Wireless Brochure, showing the Bridgwater Station.

Another fascinating insight into the operation of the Bridgwater and Bodmin Wireless Stations was published in the *'Western Morning News'* on 20th February 1936, which is worth reproducing in full:

"A small pen writes in a Bodmin wireless station. Six thousand miles away, in Cape Town, a similar pen writes precisely the same message. There is no tangible link between them, but an important message has travelled

6,000 miles in a fraction of a second at the dictate of an operator in London.

To me, that was the greatest marvel in the beam stations at Bodmin and Bridgwater of Cable and Wireless Limited. Even in this enlightened age it seems a little short of miraculous that a mechanical pen on the other side of the world can be made to imitate, word for word and letter for letter, what its twin brother in Cornwall is writing.

I spent two interesting days in the two West Country beam stations. My first experience was with the outgoing part of the business at Bodmin. That station is virtually the brains of the system, controlling the hands of a robot writer in Montreal or Cape Town. Steel arms reach up to 300 feet into the sky over Bodmin Moor, but the hands and fingers are 6,000 miles away, responding to the decrees a little machine in the transmitting house.

It was the impressive aerial system, more than a mile in length, which first attracted my attention. Wires suspended between them appeared like a gigantic cobweb, one section pointing towards Africa and the other to Canada. These are the beam aerials, so efficient that only a fraction of the power normally used by telephony transmitters is required to give reliable communication across the world.

Inside the main room of the transmitting-house, conversation was impossible in the din of swift-moving machinery. Engines and generators set the floor pulsating. It was a strange contrast to the outside quiet.

A sound-proof door led to the transmitter-room, a place of glowing valves and complicated instruments which ticked out an unintelligible stream of Morse messages. They were incomprehensible for two reasons.

It is impossible to read Morse at a speed of 200 words a minute, and still more impossible to read a message which has been turned upside down, metaphorically speaking, for its journey across two Continents.

Occasionally the engineers switch in a small machine which bears a remote resemblance to a recording barometer. It intercepts the message on its way to the transmitter and aerials and gives on a moving strip of paper a pictorial impression what is about to be projected into the ether.

Almost at the same instant, a similar machine is working in Cape Town or Montreal. If the pen at Bodmin inscribes the letter' A' in Morse code on the strip, one thirtieth of a second later the pen in the Cape Town receiving station has written 'A' on the fast-moving tape. The whole process is accomplished at an incredible pace. By speeding up the motors at the transmitting and receiving ends, speeds of five and six hundred words a minute have been reached. It is customary, however, to work at a steady 200 words per minute.

It has been stated that the Germans employed rather a similar method for communication during the Great War. They made a gramophone record of the vital message. Then they transmitted that recording over landline with the gramophone motor working as fast as possible, so that the message would be high-pitched and unintelligible if intercepted. By reversing the process at the other end and slowing down the reproducing instrument, they were able to make it intelligible.

The filaments of the transmitting valves are as hot and nearly as large as the electric radiators in domestic use, but a cool stream of paraffin oil, flowing around the anodes of the valves, carries away superfluous heat. This cold douche on the vital part of the equipment is of great importance, for the transmitter must always be cool and efficient carrying out its duties. It has no rest. It works day and night and is the main link in a chain covering six thousand miles.

Except for the necessary supervision and maintenance, the station is automatic. It accepts the messages from London and passes them on over ether to Montreal and Cape Town.

So great is the demand upon the company's services that a new system which enables two messages to be handled simultaneously has been pressed into operation. By means of two audible tones, two distinct communications are sent concurrently over the same line to Bodmin. There they are separated by filters and passed on for transmission. The transmitters, which operate normally on wavelengths between sixteen and thirty metres, are waiting for the messages and project them into space at a steep angle from the beam aerials the appropriate directions.

Hidden unostentatiously in a corner of the transmitting-room is a small box. It is a wireless receiver picking up transmissions from the aerials overhead.

The engineers can connect it to the telegraph printer and see that the message going out into the ether is satisfactory. For all practical purposes the signal reaches Cape Town as quickly as it travels the 100 yards to the little receiver.

At the Bridgwater receiving station I saw a wireless telegraph message which had made an unexpected trip around the world.

A Japanese station transmitted the message. It was received during some experiments at Bridgwater, and a curious repetition of certain parts of the message was observed on the printed tape passing through the machine.

It proved to an 'echo' of the original signal, which had made a complete circuit of the world and arrived one-seventh of second later.

In this station the real fascination of long-distance communication can be experienced. The telegraph printer which I saw at work was actually deriving impulses of energy from a transmitter 6,000 miles away. It might have been printing a message from a South African jewel company to the London office or an order from a British industrialist intended for a concern in Cape Town.

Signals reaching Bridgwater from those widely-separated communication posts in two continents are passed through tuned copper tubes to the receiving house. This building is by no means as imposing as its companion at Bodmin, but it is a hive of scientific research and wireless activity.

The most elaborate precautions are taken to ensure that no fraction of the incoming message is lost. There are even arrangements for overcoming that bugbear of all wireless communication - fading.

In London, a special department makes a study of wireless conditions from the astronomical standpoint. 'Sun-spots' and other phenomena are examined. It can be predicted with some certainty when reception conditions will be bad. A message is accordingly sent to all transmitting and receiving stations. Communications are duplicated by a mechanical process during transmission. A similar procedure amalgamates the duplicated parts of the message at the receiving end.

It is highly improbable that both parts of the repeated message will lost through fading. The signals are stored in condensers connected successively by a quickly-rotating arm. They are released with the appropriate interval to the telegraph printer. The result is that if the printed message is deficient in a certain place, that gap will be filled by the repeated passage in due course.

Sometimes, owing to chaotic atmospheric conditions, the waveform of the received signal is completely distorted. Engineers at Bridgwater are able to 'doctor' the signal so that the message is delivered properly to London. This wireless chain across the world is truly a silent service. It has greatly progressed since that historic day when Marchese Marconi established communication between Newfoundland and Cornwall. In its present state much of the unfailing efficiency of the system is due to the work of the technical staff, who are directed at Bodmin by Mr. R. N.

Barrington (engineer-in-charge) and Mr. D. Wilson-Jones (assistant engineer), and at Bridgwater by Mr. P. C. Fox."

One of the great advantages of the 'Beam' system was that it was extremely difficult to decode the high-speed transmissions without specialist equipment. A fascinating article written under the *nom-de-plume* 'Detector' in the early 1930s gives a detailed account of how secrecy in the transmissions was obtained:

"Extraordinary improvements in the Beam Wireless telegraph system have taken place since 1931 when the Eastern Telegraph Co. and the Wireless Telegraph Co. of South Africa became merged as the Overseas Communications Co.

By adapting for the beam the principle of the patent regenerator system used by the cables, the Company is now operating between Cape Town and London a wireless service that ensures complete secrecy at all times, and one that is completely mechanical in all its phases of working.

Before the merger of the former rival interests, there was a common factor – distortion – that made its presence felt when operating at high speed on both the cable and the beam. Wireless was always more speedy than the cables, but distortion crept into both systems of working. The result was that while the beam might be transmitting a message from London to Cape Town at 200 words per minute, the signals, as reproduced on a tape at the receiving end, although they could be ready with accuracy by sight, were invariably not sufficiently perfect, due to distortion, to be fed into a printing machine and reproduced as English at the same high speed. Distortion due to another cause seriously limited the perfection, at speed, of the cables.

Shortly before the companies merged, however, the cable engineers perfected their regenerator system, which, briefly speaking, put an end to this limitation.

Soon after the merger, the engineers of both companies began experiments to adapt the principle of the direct current regenerator system to the needs of the high frequency oscillatory current as used by the Beam.

These experiments have been completely successful and for many months past the South African Beam has operated what is known as a double current cable code system which is unknown anywhere else in the world.

Beam Wireless and submarine cables terminate in the same room at Kodak House, Shortmarket Street, Cape Town. Today the two systems are complementary and supplementary to each other, which, in itself, has made for remarkably increased efficiency in the realm of world-wide communications from South Africa.

But the introduction of the double current cable code system to the beam outshines any other improvement resulting from the merger.

Formerly, one tape transmitting machine, located on a table in Kodak House, operated the beam transmitting station at Klipheuvel. Today, two transmitting machines, side by side on the same table, are used simultaneously.

Each machine deals with its own message and has a 'channel' of its own until Klipheuvel is reached. Here the two messages are jumbled into one and shot over the beam, to be automatically dissected as two messages again on the other side.

While travelling between Klipheuvel and Bridgwater at the speed of light, the two messages are fused together and are completely unintelligible to the ordinary wireless listener. And, assuming that some clever person manufactured a receiver capable of recording automatically the beam signals at anything up to 200 words per minute, the result would still be useless because the individual Morse signals of the two messages are jumbled to such an extent that the result would be completely meaningless.

One hundred per cent secrecy therefore, has been achieved, the importance of which, for many reasons, cannot be over-estimated.

To the Company, however, the great importance of this invention, at the moment, lies in the fact that the almost limitless speed of the radio which was formerly confined to the ether between the transmitting and receiving aerials, can now be extended to the mechanical realm beyond the point of reception.

In other words, it is now possible to transmit directional wireless signals with guaranteed secrecy and to have them reproduced automatically at high speed into English or any other language at the receiving end. Moreover, if a message is intended for an inland destination, it is possible to use the perforated tape, as received, to operate a land-line transmitter without any loss of time.

The position is that with this invention, and the many other mutual improvements resulting from the union of the beam with the cables, the Overseas Communications Co. of S.A. Ltd., is now in a position to handle without congestion in any one day, twice the volume of traffic that, as yet, has ever come to Kodak House within 24 hours."

The Engineering Department of the General Post Office inspected the Bridgwater Station during 1934, and their report appeared to be (on the whole) very complimentary:

Masts and Antenna Systems

Considerable maintenance work and re-arrangement has been carried out on the antenna systems. 16, 22, 32 and 60 metre antennas are available for the Canadian Service, whilst 16 and 33 metre antennas are available for the South African Service.

The extreme masts of each four-mast system have been fitted with lanterns at the mast heads, in accordance with Air Ministry requirements. In each case the usual associated lighting standard with lantern, has been erected near the mast base. Arrangements are in hand for masts and associated ground iron work, including antenna feeders, to be repainted at an early date.

Receivers

The two simplex and two multiple type receivers are giving satisfactory service and appear to be in good condition.

Power and Battery Room

The Aster oil engines, the use of which was superseded prior to the last inspection, are run-up monthly for maintenance purposes. The cooling tanks, for engine cylinder circulating water, together with associated pipe work will be repainted in the near future.

The induction motors, D.C. generators and switchboards are in a satisfactory condition. Some of the positive plates of the 8 volt, 720 ampere-hour Tudor filament battery are buckled and the battery is stated to be exhibiting non-retention of charge symptoms. The matter is receiving attention.

Buildings

The weather sides of the station buildings were treated some six months ago with 3 coats of Szerelmey Stone liquid. The Staff houses have been completely treated in a similar manner. The painting of inside woodwork, cast iron guttering and piping etc. is due to be carried out. Arrangements are in hand for this to be done.

The wooden and galvanized iron shed which has been erected on the site is proving invaluable for the carrying out of aerial maintenance.

Land

The grass on the site is cut regularly. The boundary fences are being maintained in good condition. The surface of the road on the site running past the Staff houses and leading to the Station buildings is slightly holed in places and will shortly be made up.

General Observations

The general condition of the buildings and plant at Bridgwater is satisfactory. Maintenance work is being carried out in accordance with the terms of the lease, which are appreciated and respected.

Mr. Allen, the Acting Engineer-in-Charge at Bridgwater very courteously facilitated the inspection.

There was an interesting development in early 1937 between the Air Ministry and Cable and Wireless Ltd. with regard to the lighting of masts at the various wireless stations in the UK, including Bridgwater. A letter from Mr. A. H. Self of the Air Ministry stated:

"I am commanded by the Air Council to refer to Air Ministry letter 25946/30/S.6 of the 31st August 1932 addressed to Messrs. Imperial and International Communications Ltd., and to later correspondence relating to the lighting of W/T masts which constitute a danger to air navigation, and to inform you that owing to the continued increase in night flying they now consider it necessary for these lights to remain switched on or lighted from sunset to sunrise.

The Council are anxious for this new arrangement to come into force at the earliest possible date and would accordingly be glad to learn whether your Company will be prepared to issue instructions for this revised procedure to be adopted forthwith at Brentwood, Bridgwater, and Grimsby (Tetney). It is understood that the obstruction at Ongar already remains lighted throughout the hours of darkness.

I am further to suggest that as these obstruction lights are also lit during periods of bad visibility during the day, the question of lighting control by sun valves rather than by hand switches may receive early consideration."

Cable and Wireless responded on the 18th February 1937:

"We have to refer to the provision contained in Clause 6 of the agreement, dated the 11th February 1935, between the President of the Air Council and this Company regarding the lighting at night of the aerial masts at the Bridgwater Wireless Station.

In this instance it is provided that advice of any failure of the lights should be immediately telegraphed to the Air Ministry and it occurs to us, in view of the extension of hours of mast lighting which has recently been adopted at this and certain other of the Company's Wireless Stations, coupled with the inaccessibility at night of Post Office facilities, especially at Tetney, that a quicker method of communicating the information in question should be devised.

We would submit for your consideration that, in future, advices of light failure affecting any of the four stations concerned should be telegraphed to our Central Telegraph Station at Moorgate over the Company's own system, thence communicated to the Air Ministry by telephone. Subject, therefore, to your being in agreement with this proposed amended procedure, we should be glad to issue the relative instructions, if you would kindly advise us of the telephone number of the Official to whom you desire such information to be communicated."

Boxing Day 1937 saw the station being put out of operation for a short time when four local youths managed to interrupt the main electricity supply to the station by throwing a stick towards a small sub-station in Bridgwater and shorting out the main circuitry, potentially causing severe damage. The youths were bound over in the sum of £5 to be of good behaviour for 12 months, and had to pay £5 6s in costs.

The Imperial Telegraph Bill was still being processed in Parliament, and a report which appeared on 31st May 1938 gave a brief overview of the new Bill:

"Parliamentary consent was given tonight on a second reading by 186 votes to 110 to the Imperial Telegraph Bill, which gives effect to the agreement between the Government and Cables and Wireless, Ltd.

Captain Euan Wallace, Financial Secretary to the Treasury, explained that the Bill gave legislative authority for a substantial reduction of Empire telegraph rates, which had already taken place, and relieved the present strain on the finances of Cables and Wireless, Ltd., cancelling the rental payable to the Postmaster-General for the four Beam Wireless stations in return for the acquisition of a share in the equity of the whole undertaking. The life of the licence granted would be extended until 1963. It was essential, he said, if we were to have an Imperial communications system to the best advantage of the consumer that we should have some form of unified arrangement which commanded the assent of all the Governments which were partners in this concern. For that and other reasons the Government remained convinced that the most efficient service would be obtained by a company working under these quasi-public utility companies. The Socialists opposed the Bill on the ground that public utility was sacrificed to private gain by the disposal of valuable State undertakings to private interests"

Unsurprisingly there was a degree of local opposition to the Bill, and a great deal of uncertainty on the future of staff currently employed at these stations.

Liberal M.P. for Barnstaple, Mr. Richard Acland, who seemed to be a person on a mission to defeat the Act, wrote a heartfelt letter to the North Devon Times on 31st May 1938:

Dear Sir,

I have dealt, by now, as every Member of Parliament is bound to deal, with a fairly large number of claims from constituents against the government for widow's pensions, war disability pensions, unemployment benefit and the like. As a result, it is my very definite impression that a general instruction has been passed round the departments that in these difficult times, when every penny is needed for increased armaments, the departments are to use every possible rule of law and of regulation to save the taxpayer's money. A quite different set of rules, however, seem to apply when it is a question of meeting the much less well-founded claims of 'big money'; and we had an example of this governmental generosity in the House yesterday. I do not suppose one in ten thousand electors and taxpayers is even aware of the Imperial Telegraphs Bill which passed its second reading by 188 votes to 110 after less than 20 of those who voted for it had heard the discussion that took place.

Cable and Wireless Holdings Company, with its eyes open, made an agreement with the government in 1928 whereby they leased, for 25 years, the Government's Beam Wireless stations (including those at Bridgwater and Bodmin - well-known to West countrymen) for £250,000 a year. In 15 years' time, therefore, those profitable means of communication would once again have become the unqualified property of the nation - that to say, of the taxpayers. Communication by cable is distinctly less profitable than communication by wireless, and, recently, the company has been meeting foreign competition, and, as a matter of ordinary business, has reduced its of charge to a universal flat rate, as a result of which the company expects, after a period of reduced revenue, to increase its business and increase its profits - much as the Post Office ultimately increased its revenue out of the reduced flat rate of ls. for all calls after seven p.m.

No doubt it is a very good thing for all concerned that the company should in this way reduce its rates.

But the occasion has been used by the Government to consent, and to persuade its willing majority to consent, to a bargain which has everything possible to recommend it from the point of view of the company, but is sheer dead loss as far as we who are taxpayers are concerned.

We have not only cancelled our right to resume possession of the Beam Wireless stations in fifteen years, we have also released the company from its obligation to pay £250,000 per year rent until then. It is impossible to set any cash value on what have given up under the first of these heads, and £250,000 year for 15 years, is today worth £2.875,000. We receive, in return for both these concessions, a parcel of shares in the company worth today £1,100,000.

The Government spokesmen tried to make a suggestion that somehow or other this rotten bargain was a necessary part of the scheme by which, the company reduced its rate. But spite of question and answer, I could not see that there was any connection between the two whatever.

I put the point in debate - which was subsequently taken up by Mr. Pethwick-Lawrence speaking from the Labour front bench, and was not answered by the Government - that if the Government had been the directors of a company, instead of the trustees for the taxpayers, and if we in the House had been shareholders in the company, we would under no circumstances whatever have sanctioned our directors in making such a miserable bargain. And yet, Conservative members, obeying their chief whip, put the bargain through without a single murmur.

And so sir, next time you pass Bodmin or Bridgwater wireless station, you will perhaps remember that that station was once yours, but today, unless you happen to own shares in Cable and Wireless Holdings, your property has been given away to a private company for a song.

If it had been argued in the House that the Company was in difficulties with foreign competition and needed assistance, then we might have listened to the arguments, and in these days we might have granted a subsidy of such a size and for such number of years as might have seemed necessary. But just to give away valuable national assets for all time, far below even their present value does not seem to me to square with the attitude of the Government, towards the war pensioner, the widow, and the unemployed man.

I said in the course of a short speech on the subject, and I think it is true, that the Liberal Party aims at creating the conditions which any man who has enterprise will receive his reward whether he has capital or not. It is an intensely difficult aim to achieve and no one knows better than how far we are from having achieved it today. But there seems be growing up a rather different point of view among Government supporters, namely that be who has capital - and even he who has watered capital - must be assured of his reward whether he has enterprise or not. This seems to me to be wrong.

Yours sincerely,

Richard Acland, House of Commons, 31st May, 1938.

In June of that year, Mr. Acland brought the matter up in Parliament:

"Mr. Richard Acland M.P, in the House of Commons on Monday, with other members of the opposition, made a strong protest against handing State property to a private concern. The Imperial Telegraphs Bill was being debated on the second reading. As the result of the Bill, the freehold of the Bodmin and Bridgwater Beam Wireless Stations and two other beam stations will be transferred from the Post Office to Cable and Wireless, Ltd., and the rental of £250,000 a year will be cancelled.

In return, the Government will receive 2,600,000 £1 shares in Cable and Wireless, Ltd., which, on the basis of the 3½ per cent dividend paid last year, would bring £91,000 a year.

Mr. Acland asked what connection there was between the reduction in cable rates and the bargain made about the four Beam Wireless stations, which would result in a loss of £100,000 a year or more to the Government. As he saw it, this company was losing trade, and with the object of regaining it and making higher profits, it had reduced its rates. Usually when a company made losses it went out of business. They were told this company was vital, and could not allow it to go out of business, but surely, if that were so, it would be better that these services which could not be made to pay by private enterprise should become a State-controlled combine. To make a bad bargain, which involved the State in a loss of £100,000 a year, did not seem a reasonable way to deal with it at all.

The Liberal principle was that where there was a service which must be a monopoly there must be control. He did not believe the Government would have sufficient control in this case. If the Government had been in the position of the directors of a private company and had made as bad a bargain as this, they would have been dismissed by their shareholders at the next meeting. It would have been better to-make a thorough-going job, and take the lot as a public utility concern, to be run as a national undertaking, which, incidentally, could have made profits like the Post Office, did."

The Imperial Telegraphs Act was eventually formally passed, which in effect formally transferred the ownership of the Beam Station network to Cable and Wireless Ltd., and the 25-year lease of the stations from the Post Office cancelled. The initial impact of this act was for the new owners to make sweeping changes across the network, closing the Bodmin/Bridgwater stations and replacing them with stations at Dorchester and Somerton. The Grimsby site was also scheduled to close, being replaced with a new transmitter site located at Ongar, in Essex.

The debate over the new Act makes for interesting reading, and a relevant extract gives full details of the arrangement on which the Act was based:

"Cables and Wireless Limited was set up as an operating company to conduct the main overseas telegraphic services of the British Commonwealth. The capital of the company was fixed at £30,000,000, all in ordinary shares. An agreement was made between the company and His Majesty's Government in May, 1929, but operating as from

1st April, 1928, which can be summarised under five main heads. First, the Postmaster-General leased to the company the four Post Office Beam Wireless stations situated at Skegness, Bodmin, Bridgwater and Grimsby respectively.

They were leased for 25 years at a yearly rent of £250,000 plus 12 per cent of any surplus profits which Cable and Wireless Limited, might earn over a standard revenue which was then fixed at £1,865,000, representing 6 per cent on the capital of the company. Secondly, the Postmaster-General gave to Cable and Wireless Limited, the free use of the necessary telegraphic circuits for working those Beam Wireless stations. Thirdly, and on the other side, the company paid the Governments concerned a sum of £1,267,000 for the cables which were Government-owned.

Fourthly, a number of measures of Government control were imposed on Cable and Wireless Limited. That is essential to the whole agreement, because a main point of the negotiations has been to conserve intact and in working order the Imperial cable system which, in certain circumstances, is not remunerative, but which it is very essential to keep in working order against emergencies. The first item in the measures of Government control to which Cable and Wireless Limited, submitted under the agreement was an obligation to maintain the cables even if they were unremunerative.

The second was that the chairman of Cable and Wireless Limited, was to be approved by His Majesty's Government. Thirdly, there was, and is, power to the Government to take over the whole of the plant and staff of Cable and Wireless Limited, in the event of war. There was then a provision to ensure that control should remain in British hands, and the Imperial Communications Advisory Committee, which is a body upon which all the Governments concerned in this great enterprise are represented, was given certain special powers over the company; for instance the company is not allowed to increase rates without the prior sanction of the Advisory Committee, nor may they dispose of any of their assets; and the Advisory Committee has to be consulted on any changes of policy. I put all those items under the fourth heading - measures of Government control.

The fifth and last main item of the agreement was that one-half of any surplus revenue over the standard (plus the 12 per cent which was to go to the Government straight away) was to be applied either to the reduction of rates or to the development of services. The company was, therefore, not able under that agreement to derive any direct advantage from increasing its business except to the extent of half of the surplus over their standard revenue of 6 per cent on the capital, after paying our 12 per cent of that surplus. The object of this agreement, as I have already indicated, was to provide for co-operation among the various Governments of the British Commonwealth of Nations, to maintain and develop under British control the overseas cable and wireless system of the Empire, operating under private enterprise in the United Kingdom, but under conditions which I think can be accurately described as quasi-public-utility.

During the years immediately subsequent to that agreement, various events occurred in the world which made the world not a very good place for cable companies. Cable companies depend very directly for their prosperity or otherwise upon the flow of international communications and international trade. We have to face the fact that since Cable and Wireless Limited, was created it has been unable to earn the revenue which everybody expected it would earn, and in fact it has never earned even the standard revenue of 6 per cent. The average dividend paid between the years 1930 and 1935 was 1.7 per cent, and in 1936 the company succeeded in paying a dividend of 2½ per cent. The result of that was that, so far as the Government and the public were concerned, there was, so to speak, no surplus cash in the kitty. The telegraph rates could not be reduced as much as was hoped, although in point of fact certain reductions were made. As the telegraph rates could not be reduced, there loomed up on the horizon the possible threat of foreign competition on Empire routes by the use of direct wireless telegraphy.

One consequence of the Act was a review of the Beam Wireless network. On 12th December 1938, the fate of the Bridgwater station was confirmed in Parliament during questions between Mr. J. R. Rathbone, M.P. for South-East Cornwall and Major Tryon, Postmaster General:

Mr. Rathbone asked the Postmaster-General upon what date the Post Office beam stations will, in accordance with the Imperial Telegraphs Act, 1938, be transferred to Cable and Wireless, Limited; how many and which of these stations will thereupon be closed down; and what arrangements are being made for alternative employment for those now working at these stations?

Major Tryon: Under the deed of conveyance which is on the point of being completed, the transfer of these stations will take effect as from 1st March 1938. I am informed by the company that, as at present arranged, the installations at Bodmin, Bridgwater and Grimsby will in due course be transferred to existing stations at Dorchester, Somerton and Ongar, respectively. I am assured by the company that all members of the staff affected will be transferred to stations operated by the company, with the exception of certain unestablished labourers. In these cases, I am informed that long notice and adequate compensation will be given.

It was reported in December 1938 that very little opposition to the closure of the station had been received, and it was reported that:

"Closing down of the Beam Wireless stations at Bodmin and Bridgwater under the agreement with Cable and Wireless, Ltd., has not so far resulted in any protests being made to the M.P.s for the constituencies concerned - Mr. J. Rathbone and Mr. V. Bartlett. Mr. Rathbone is well satisfied with the statement he received from the Postmaster General today to the effect that alternative employment is being found for all those now working at the stations, with the exception of certain labourers who will be compensated."

Although not directly linked to the Bridgwater Beam Wireless Station, it is interesting to note that a wireless direction-finding station was established at Stockland Bristol, about 5 miles north-west of Bridgwater, in 1939. This confirmed the suitability of the area for radio communication, and the station (combined with a stations at Wymondham, Norfolk, and Cupar in Scotland) was used throughout World War 2 to assist in locating and identifying transmissions from the continent.

A letter from H.M. Treasury dated 27th February 1940 confirmed that the closure of the Bridgwater site had been approved, along with the closure of the associated stations at Bodmin, Grimsby and Skegness. There had been some concern that the sites might have been required for use by the military authorities, but this proved not to have been the case, as the letter from the Treasury to Cable and Wireless confirmed:

"We had a discussion last Thursday at which Sir Campbell Stuart was kind enough to assist, on the subject of your claim to a credit of about £15,000 a year in respect of your scheme of concentration of stations which had not been carried into effect.

I was able to tell you that since our last talk I had consulted the appropriate military authorities as to whether, on strategical grounds, it was still necessary to ask you not to proceed with your concentration scheme and I had received the advice – welcome from your point of view – that there is no longer any objection to your closing Bodmin and Bridgwater stations and transferring some of the plant and masts to Dorchester and Somerton respectively, and to your closing Grimsby and Skegness stations and transferring the plant at those stations to Ongar and Brentwood. The way is, therefore, clear for you to proceed as soon as you care to with this concentration.

You explained again that you had first raised the question of concentration with the Government as early as December 1938 and that you are entitled to some form of compensation for the non-realisation of economies which you would otherwise had effected, You were prepared to waive the question of staff and other economies provided you could receive credit in respect of certain landline rentals. The sum at issue in this respect is the amount of the rental at current tariff rates for the circuits to Bodmin, Bridgwater, Grimsby and Skegness, amounting to £15,360 per annum.

After a detailed discussion of your claim, I said that I would be prepared to recommend as a compromise that you should receive compensation amounting to six months' rental of the landlines in question, namely £7,680 and you on your part said that you would be prepared to accept such a compromise arrangement, although in your view it did not do full justice to your position.

The payment of this amount can be best effected, I think, by deducting it from your next quarter's payment to the Post Office.

As regards the future, the Post Office will, in the special circumstances allow credit, in respect of the surrender of landline circuits to the Beam Stations, to operate from the date on which each circuit is given up without requiring three months' prior notice as prescribed by Clause 7 of the Supplemental Agreement of 25th April 1939.

But you will no doubt notify the Post Office as long as possible in advance of the dates on which you will be able to relinquish the circuits."

The process of decommissioning the site continued into the early 1940, and a letter from Cable and Wireless to the General Post Office dated 16th May 1940 requesting the removal of landlines to and from the Bridgwater site was despatched:

"With reference to our letter dated 25th April 1940, will you kindly arrange for the recovery of the following circuits as from tomorrow, 17th May 1940:

One telephone circuit between Electra House, Moorgate, and the Bridgwater Station.

Two telegraph circuits from Electra House, Moorgate, to the Bridgwater Station and thence to the Bodmin Station.

With regard to the circuit with the Bridgwater Station to be retained by use, we particularly desire the use of the line No. RA 19870 and we should be glad if you would kindly arrange this for us.

Will you be good enough to advise us, at your convenience, the effect of these changes on the rentals paid by us on account of landlines?"

In 1941, the station came close to being resurrected as part of the Government's emergency communications contingency planning; a back-up system was being established to ensure communications should international cable networks be damaged by enemy action, and one possibility considered was the installation of transmitters at the Bridgwater site.

The whole episode was fraught with political disputes and differences of opinion between the Post Office and

Cable and Wireless Ltd., and today it is fascinating to follow the story as it developed during the year.

The first mention of this possibility came during a meeting of representatives of the Post Office and Cable and Wireless, held on the 28th July 1941, to discuss the 'disposal of telegraph traffic in the event of cable interruption'.

It was recorded that:

"It has been suggested that that two transmitters may be installed at the Cable and Wireless Ltd. Bridgwater receiving station which is not at present being used. The Bridgwater station offers the advantages that the building and complete beam aerials directed on North America are immediately available. A higher grade service is to be expected than could be provided from simple aerials at the London and Scottish emergency stations. The employment of Bridgwater would also afford an opportunity of dispersing the emergency transmitters on three instead of two sites. On the other hand, control lines from London to Bridgwater would be more liable to interruption on account of their greater length and routing, and more difficult to replace than control lines to a station only 20 miles or so from London.

It is suggested that the question be considered forthwith whether an immediate decision is taken to utilise Bridgwater when the transmitters are available or whether such a decision should be deferred until the transmitters are actually available."

A letter from Mr. J. Innes, Director of Telecommunications at the General Post Office, and Sir

Campbell Stuart of the Imperial Communications Advisory Committee on 7th August 1941 gave further impetus to this idea. The 'Wilshaw' quoted in his letter refers to Sir Edward Wilshaw, Chairman of Cable and Wireless Ltd.

"Wilshaw's people hold the view very strongly that Bridgwater should be utilised as a station for emergency communications and I am quite willing to recommend that two transmitters be provided at this point from the eight which are being obtained so that Wilshaw can operate two additional transmitters in emergency."

Mr. L. V. Lewis of the Telecommunications Department of the General Post Office, also commented on the idea in internal correspondence with Mr. Innes, emphasising that Cable and Wireless ('The Company') were keen to proceed on this basis:

"The Company's representatives were most helpful and accommodating throughout the discussion and it was only on the Bridgwater question that agreement presented difficulty. It is obvious that the Company are anxious for transmitters of the national reserve to be installed at Bridgwater and that they would like to have directions from the Government to keep the station available for the purpose.

The Post Office had hoped to satisfy the Company by agreeing to consider that matter later, so that the onus of maintaining the station in the meantime would rest with the Company. The following paragraph was included in the first draft of the report:

The best course might be to defer consideration of the use of the Bridgwater station until the end of the year (assuming that the station is still available then), when it may be possible to assess more clearly than at present the advantages or otherwise of locating transmitters there. In the meantime the Wireless Telegraphy Board might be approached with a view to ascertaining whether there are any objections to the installations of transmitters at Bridgwater.

This however, was not accepted by the Company and a revised paragraph, drafted by Mr. Denison-Pender, in which consideration forthwith is suggested, was adopted.

The committee recognised, of course, that although the Company is interested financially, it is only right that the Bridgwater scheme should be considered on its merits. The output of the emergency transmitters installed in the North West of London will be limited by the simple nature of its aerials, and it is obvious that Bridgwater would have a much greater output. The vulnerability of the Bridgwater station and its control lines are of course important factors.

Informal enquiry was made of Col. Lycett of the Wireless Telegraphy Board to ascertain whether he knew of any overriding considerations which would enable the Bridgwater proposal to be dismissed out of hand. He saw no special objection to Bridgwater, however, and thought that the Service Departments would be prepared to express a considered opinion at short notice if asked to do so. The general question of Bridgwater is however being developed in a separate memorandum which the Engineer-in-Chief is preparing."

Sir Edward Wilshaw (1879-1968).

The Post Office Engineering Department did indeed produce a memorandum which was released on 8th August 1941, in which they concluded:

"During the discussion on the further facilities which will become available during 1942, a suggestion was made by Cable and Wireless Ltd. to install two of the 'additional 8 transmitters' at Bridgwater and the relative advantages and disadvantages of this proposal have been briefly outlined in Section 7 of the report. There is no doubt that a higher grade service would be obtained from Bridgwater than from the more modest N.W. London proposal but this latter proposal has to cover the dire emergency of London being cut off from the main transmitting stations.

If transmitters at Bridgwater can be considered a relief to the N.W. London equipment, no reason is seen why the existing transmitters at Dorchester and Ongar should not first be used in this manner unless they have already been put out of commission. If this latter possibility has to be covered then it would seem wise to consider installing one or two of the new transmitters Cable and Wireless are obtaining towards the end of this year at Bridgwater rather than Dorchester or Ongar as at present proposed by the Company so as to effect the earliest possible cover. Thus the Bridgwater proposal seems to this Department proper to be considered as a reserve station to Dorchester and Ongar. The proposal then becomes similar in some respects to the Post Office reserve of Criggion for Rugby v. Leafield.

A lead in the direction was given to the Company's representatives but they only pursued it as a reserve to the N.W. London scheme, whether they overlooked the wider issue or deliberately avoided it was not apparent."

Sir Campbell Stuart, Chairman of the Imperial Communications Advisory Committee, was advised accordingly in a communication of 9th August 1941:

"Since our talk of Thursday inst. about the emergency wireless services, a further point has occurred to me in regard of the desire of Cable and Wireless Ltd. that certain of the eight transmitters of the national reserve should be installed at Bridgwater.

We fully appreciate, of course, that their Bridgwater proposal has the advantage that buildings and aerials are already available, but the best solution would seem to be for the Company to install at Bridgwater one or two of the four transmitters which they are obtaining for themselves before the end of the year. If this arrangement was adopted there would be an early reserve against possible damage at Ongar or Dorchester and at the same time we should ultimately be able to install four of the national transmitters at each of our four emergency stations in London and Scotland.

You will perhaps keep this view in mind when we come to discuss the subject with Wilshaw."

Sir Edward Wilshaw duly responded to the Post Office's Director of Communication, Mr. J. Innes, on the 20th August 1941:

"With reference to our conversation yesterday regarding the suggestion to install at Bridgwater the SWB8 transmitters ordered for Ongar and Dorchester, I find that these transmitters are being supplied without the Main and Auxiliary power supply equipment, as suitable supplies are available at both the latter stations.

If the proposal to install this plant at Bridgwater is proceeded with, it would necessitate ordering further power supply equipment and I am afraid delivery of this would be delayed unless we could be given a high priority."

A further communication on this matter from Sir Edward Wilshaw was sent on 2nd September 1941:

"Referring to our recent conversation, I have been looking into the possibility of rapidly erecting two transmitters on our Bridgwater site for use in the case of cable interruption.

As you know we have, amongst others, two SWB8 transmitters on order for Ongar. Delivery of these is promised in November but they are not complete as they are intended for use with old 15 kW transmitters to enable the latter to be used on more than one wavelength so they been ordered without high tension equipment which they will obtain from the station's supply.

The Marconi Company informs me that they could earmark H.T. rectifiers for these transmitters, and so make them available for Bridgwater, and still keep their original delivery date of November.

This however seems to be robbing Peter to pay Paul because, if these transmitters are diverted from Ongar, the increase in our equipment there, visualised in our recent meeting with the G.P.O. officials, will not be available until they are replaced.

The Marconi Company hopes to increase their manufacturing plant in the next few months which they say might enable them to deliver replacements during the first quarter of next year, but normally they could not do so until the completion of their present orders, which will not be until this time next year. I am afraid therefore that we must look upon this replacement date as being very uncertain.

I understand, however, that they have something like 81 SWB8 transmitters on order on priorities which are as high or higher than G.P.O. priority. I do not know of course how important the purposes are for which these are required but it seems to me that with such a large number, two might be found whose purpose is not as important as the one we have in view. Would it be impossible to investigate this point because if two of these transmitters could be diverted to us, an emergency station at Brentwood could be arranged in a very short time?"

On receipt of this communication, an internal memorandum dated 11th September 1941 between Messrs. Innes and Lewis of the Post Office seemed to cast doubt on the intentions of resurrecting the Bridgwater station by Cable and Wireless:

"It appears to us that Sir Edward is trying to place objections in the way of installing two transmitters at Bridgwater. If, as would appear on his own showing, there is no insuperable difficulty to early installation at Bridgwater, the other question which he mentions would not seem to arise."

Mr. Innes formally replied to Sir Edward the following day:

"With reference of your letter of 2nd September to which I am sorry I have not replied sooner, I should perhaps make it clear that under my suggestion you would arrange to install at Bridgwater two of the four new transmitters which you hope to have available by the end of the year. Under this arrangement it is true that additional plant which you would otherwise have installed at your other transmitting stations would be reduced accordingly, but I can hardly agree that this would be 'robbing Peter to pay Paul', since it would offer the advantages, which you have yourself stressed, of spreading plant over three, instead of two stations, and of making use of buildings and aerials already at your disposal at Bridgwater. I note that the Marconi Company could earmark the necessary H.T. rectifiers for these transmitters and still keep to their original delivery date so that early installation at Bridgwater would not present any technical difficulties.

If, of course, at a later date you wished to increase your plant at Ongar or Dorchester, this would be an independent matter; but the immediate question is one of an alternative allocation of the Company's plant and not a supplement to the number of transmitters considered by the joint committee."

On 16th September, Sir Edward gave an update, basically confirming that the Bridgwater plan was still being strongly considered:

"If we delay the SWB8 drive panels at Ongar we shall definitely delay making the maximum use of our high powers transmitters there and thereby delay the provision of wireless circuits to the rest of the world as distinct from North America.

These are in fact transmitters considered by the Joint Committee for handling traffic to the rest of the world, and are not merely for the benefit of the Company's services.

It seems preferable therefore that we should order the two complete transmitters for Bridgwater to be delivered at a later date, unless two can be delivered as suggested in the last paragraph of my letter dated 2nd September.

Meantime if this is agreed we could make other preparations at Bridgwater such as running new feeders, altering the interior of the building, arranging power supply etc., all of which will take quite a little time with the present labour difficulties."

It took a few weeks and some internal discussion within the Post Office before a reply was forthcoming, but on October 16th Mr. Innes replied:

"I am sorry I have not been able to send you an earlier reply to your letter of 16th September concerning the disposal of your new transmitters, but I felt it necessary to consider in detail the questions you raised, having in mind the desirability of not only of your making the maximum use of these transmitters, but of minimising the risk of loss by damage or isolation.

It was because of the risk of damage to Ongar (or Dorchester) that your proposal to make use of Bridgwater commended it to me and I suggested to you that two of your transmitters should be installed there; but in view of the power-effectiveness resulting from the dispersal of the SWB8 panels and since the aerials at Bridgwater are directional and therefore too inflexible for an effective reserve station for 'rest of the world' traffic, I now feel it would be best if you retained the original distribution of your four new transmitters and relied on the reserve capacity in your own and the Post Office system to make good any losses by enemy action.

There is the further consideration that Bridgwater is relatively near the landing point of the cables, the capture of which by the enemy would be one contingency necessitating the introduction of the emergency wireless service, and this together with the long landline connection militates somewhat heavily against the Bridgwater scheme."

Sir Edward replied 2 days later, on the 23rd October, and accepted the comments which Mr. Innes had put forward:

"Thank you for your letter of 21st October from which I see you agree that it would be best if we retain the original distribution of the SWB8 panels at Ongar and Dorchester. We will proceed with this installation of these as soon as possible in order to have the advantage of the extra capacity.

I also note from reasons of its relative nearness to the landing points of the cables and long connecting landline, you do not consider Bridgwater is a suitable station to be included in the emergency scheme.

I will therefore proceed forthwith with the dismantling of this station."

On 4th November 1941, a detailed internal Post Office memorandum confirmed the proposal to use the Bridgwater station was now no longer an option:

"At the meeting of the I.C.A.C. held on 3rd September 1940, Cable and Wireless Ltd. informed the committee (via its Chairman) that they had received notification from the Treasury that they might proceed with the concentration of the beam stations into two groups.

This seems to preclude any official objection to the dismantling of the Bridgwater station, now about to take place, but it is a regrettable step in so far as retention of the station would provide against the destruction of Somerton from the air even though it might contribute little to the North American wireless scheme which was formulated primarily owing to the concentration and relative vulnerability of the Atlantic cable heads.

Col. Lycett of the W/T Board has been informed by telephone of the abandonment of the proposal to use Bridgwater for some of the emergency transmitters, and of Sir Edward Wilshaw's expressed intention to dismantle the station. It is not proposed that the Post Office should urge the Company to leave the Bridgwater station in being; but in view of the lapse of time and the change of circumstances since the Deputy Chiefs of Staff Sub-Committee gave their assent to the closure (6th February 1940), the I.C.C. should be advised of the Company's decision.

Mr. Lewis has raised this matter in discussion with the Director of Communications. Mr. Innes said he will communicate with Sir Campbell Stuart on his return from America."

The masts and aerials were subsequently (and with great difficulty) dismantled during 1942, and the land returned to agricultural use. However, there were to be a few more twists along the way for the former station.

CHAPTER 6

THE LEGACY

Despite the aborted use of the site as a transmitting location during World War 2, the buildings remained intact for a few more years, being effectively 'mothballed' by Cable and Wireless, and the equipment had been relocated, mainly to the Somerton receiving station.

Although the station had been closed for commercial operations since 1938, a report in the local press on 2nd December 1950 raised the possibility of a new station being established on the site:

"An Admiralty scheme to make use of the old Cable and Wireless station considered letters from the County Council and Ministry of Town and Country Planning, stating that the Admiralty proposed to establish a wireless telegraphy station there.

It was intended to erect 108 masts in the immediate vicinity, underground cables and erect a new station building. For technical reasons it would be necessary to control the use of an area of land surrounding the station.

It was agreed to obtain the views of North Petherton Parish Council before making observations on the proposals, the Clerk and Engineer being instructed to meet the Parish Council."

A further article published on the 30th December 1950 reported on the progress of the scheme:

"In regard to the proposal of the Admiralty to establish a wireless station on the site of the former wireless station at Huntworth, the Clerk had consulted North Petherton Parish Council and the Plans Committee decided to make certain observations.

The Council should be satisfied that there was no other suitable and less productive agricultural land which could be used. Provision should be made for safeguarding the occupier. It was considered important that the masts and corridors concerned should be adjusted to avoid buildings at Huntworth Park and Little Bakers.

The Council were concerned as to the effect the establishment would have on the position of the locality, in regard to Civil Defence in a national emergency, especially in relation to evacuation or reception of evacuees."

It would appear that the Admiralty never took this idea further, and the site was eventually sold, with the masts and aerials having been removed. For the next thirty years or so, the site was used as part of a farm, with the buildings being converted into new units for agricultural use and storage.

Some photographs of the station buildings were taken in 1983, and show them in their new role. A selection are shown over the next few pages.

*This was the original battery room,
converted for storage use.*

*Another view of the battery room with new door fitted
and concrete cattle pens attached.*

Part of the Bridgwater battery room attached to the main building.

General view of the station building.

The main building looking the worse for wear.

The remains of one of the aerial mast guy support blocks.

1996 aerial view of the site, where the mounting marks for two of the masts can clearly be seen.

A well-known local Company, S. Notaro, formed in 1955, now operates a large factory on a part of the former receiving site, producing Windows, Conservatories, Driveways, and other related fittings. However, their plans for the site proved to be controversial.

In 1961, the owner of the company applied to build 4 bungalows and an access road on the site, but his application was refused by the local Planning Officer, leading to a local enquiry. The Minister of Local Housing upheld the refusal, stating that "the enlargement of this small and somewhat isolated hamlet should be discouraged."

Another application for construction of a house on the site by the same company was again refused in 1963, stating the same reasoning.

1987 saw a further application for expansion of part of the former radio station site submitted:

"Impressive plans for a waterways visitor centre at Huntworth, near Bridgwater, have been submitted to Sedgemoor District Council. They include a marina, wildfowl pool, restaurant, 10 holiday flatlets and an interpretation centre alongside the Bridgwater to Taunton canal. The outline application, by local builders S. Notaro Ltd., already has the support of the British Waterways Board, Inland Waterways Association and West Country Tourist Board. The canal is currently being restored with plans to spend more than £153,000 of local authority and grant money on restoration work. There is already a private 10-acre wildfowl reserve on the site and if the scheme goes ahead and this is opened to the public, the Slimbridge Wildfowl Trust will be asked to give advice. The canal mooring basin will provide berths for about 30 craft."

However, by 1989, permission was finally obtained when it was reported that:

"Sedgemoor District Council planners have given the go-ahead for a 52-bedroom hotel and waterways visitor centre at Huntworth. The complex, alongside the Bridgwater to Taunton canal, will be built on an 18-acre site currently used for agriculture, and create up to 70 new jobs.

The planning committee agreed to the application for outline planning permission from local builders S. Notaro. Councillors were told the hotel development would boost tourism and outline permission already existed for a smaller scheme.

North Petherton Town Council and seven individuals objected to the plans for the hotel, waterways visitors centre, canal mooring basin, exhibition centre and other facilities. Part of the canal has recently been reopened for navigation for the first time in decades and there are plans to restore more of the waterway.

Councillor Michael Payne (Ind, North Petherton) said it was a very dissimilar application from the original one, was far too extensive and local roads would not cope. But Cllr Mrs. Julie Hooper (Con Bridgwater Quantock) said it was exactly the sort of scheme needed and was the beginning of the redevelopment of waterways in the area. The committee agreed that the developer would have to carry out road improvements and landscaping."

However – in 1990 the scheme was sensationally thrown out by the local Planning Authority:

"A £12 million hotel and holiday homes scheme has been thrown out by Sedgemoor planners. The canal side development at Huntworth just outside Bridgwater, would have created at least 120 jobs. But the district council planning committee decided it was too big for the 14-mile long canal.

Local builders S. Notaro Ltd had backing from a financial partner for the venture. They will now consider an alternative site alongside the canal but in Taunton Deane Borough Council's area.

Company managing director Joe Notaro said: "The councillors are not forward people at all. They are turning Bridgwater into a graveyard. My financial partner is devastated."

Cllr Trevor Donaldson (Conservative, Bridgwater Quantock) accused his fellow councillors of "having grasshopper minds" and said they should look to the future.

Notaro already has permission for a 60-bedroom hotel on the site. But they claimed to make it viable, they needed to double the size and build 100 holiday flats for sale.

The canal, which is being re-opened for boating, could eventually become part of a 70-mile system of waterways around Somerset. But that will depend on a £7 million barrage being built across the River Parrett. Six local authorities and organisations have been asked to put £200,000 each into a feasibility study for the waterways scheme."

However, a plan to expand the site with a similar scheme in 1991 met local opposition:

"Villagers at Huntworth are united in their opposition to a £12-million hotel and holiday homes complex. It has been put forward by local builder S. Notaro even though the company claimed last year it would take the proposals elsewhere when a similar scheme was rejected.

The company already has planning permission for a 150-bedroom hotel alongside the restored Bridgwater to Taunton canal. It has now offered to build a new road if the hotel can be doubled in size and 100 self-catering flats built, creating up to 120 jobs.

Local residents, there are less than 100 - fear their peace will be shattered and that the flats could be sold off and the rest of the area opened up for development.

They have signed a petition against the plans and called a public meeting for later this month. The company has agreed to carry out an environmental impact assessment. It says people should be glad to see the village to be promoted as part of the inland waterways scheme. The proposals are due to go before Sedgemoor District Council planners next month."

In 2009, Lakeview Holiday Cottages, a self-catering holiday estate, was eventually established on part of the former Wireless Station site, which continues to flourish to this day.

The land of the former radio station and its surroundings continues to develop, with a large business park close to Junction 24 of the M5 now constructed, with plans to expand and develop the area even more being considered.

Although the station was only active for around 14 years, the importance of the station cannot be over-stated. It can be remembered (like the Chedzoy station) as part of the United Kingdom's radio history, and its role in developing long-distance communication must never be forgotten.

Today, the aerials are long gone, although one can establish their former location from satellite views of the area where hedgerows and ditches follow the path of former masts and wires.

The main building is still standing, although now converted into 5 private accommodation units. Their history is still remembered, however, as the postal address of these buildings is known as 'Beam Wireless Station, TA7'.

Some local business are also to be found on the site, ranging from catering to equestrian supplies.

It is still possible to walk around the area and imagine what used to be there; the quiet (almost eerie) silence, only disturbed by the noise from the nearly M5 motorway, can remind us of those pioneering days when the wind whistling through the aerial wires was on the only sound providing a background to this tranquil setting.

As a footnote, it will be of interest to briefly review the Canadian counterpart stations of the UK/Canada circuit.

The Marconi's Wireless Telegraph Co. Ltd. and its Canadian subsidiary, the Marconi Wireless Telegraph Co. of Canada, which became the Canadian Marconi Co. (CMC) in 1925, set up their transmitting station in Drummondville, Quebec and a receiving station at Yamachiche, Quebec, near Trois-Rivières.

Radio telegrams and telexes received for Canadian distribution were forwarded to the existing telecommunication companies in Montreal. The Yamachiche receiving station also functioned as a research, repair and custom radio manufacturing facility for CMC.

A recent view of the Bridgwater Beam Station building.

The wireless telegraphy service connecting London to Canada and on to Australia started in 1926. As telephone technology improved, the radio telephone service between London and Canada was added in 1932. Gradually this service was extended to other destinations. Radio facsimiles, mainly for sending wire photos, were added in the late 1930s.

After the Second World War, it was decided that Canada's international telecommunications should be handled by a government body rather than a private company. In 1950 a Canadian crown corporation, the Canadian Overseas Telecommunication Corp. (COTC), was established to take over CMC's wireless facilities. This led to new receivers being installed at Yamachiche.

Teletype and TOR (teletype on radio) circuits were installed with additional connections to Germany, France,

Italy and Bermuda. The COTC also established wireless services to South America and to Europe. By 1968 the COTC had 6 radio telephone and 9 radio teletype circuits, its maximum number.

Technological advances continued to change telecommunication methods. High capacity undersea cables, which carried multiple voice and data channels were introduced, such as the TAT 1 cable from Britain to Canada, was inaugurated in 1956. By the mid-1960s, commercial satellite communication further cut into the role of Drummondville and Yamachiche.

When the wireless service ended on 23 June 1975, only the circuit to Greenland was still functioning. In the same year COTC became Teleglobe Canada. The main Yamachiche building became a furniture manufacturing company, and in 2011 there was a serious fire which destroyed the building.

CHAPTER 7

AN OVERVIEW OF BODMIN
WIRELESS STATION

It is worth briefly considering the role of the 'associated' transmitting station to Bridgwater, namely the Bodmin site in Cornwall. Although some aspects of the station have already been covered, more detailed information is of interest. It has a fascinating history, recalled by former employee J. R. McCullum in 1977 and updated by J. D. Sharples in 1983.

"Bodmin Radio Station, perhaps the least known of the stations within the International Executive, but not the least important, is situated in the parish of Luxulyan in the County of Cornwall. Five miles from Bodmin, fifteen miles from Newquay and forty-five miles from Penzance and facing the main A30 truck road, its masts have been a familiar landmark to the hundreds of thousands of holidaymakers travelling to and out of the county by car during the holiday season.

In order to bring out the salient features of the station, it is necessary to delve into history. In July 1924, the Marconi Wireless Telegraphy Company Limited signed contracts with the British Government, the Dominion of Canada, Australia, India and South Africa, for the construction of radio stations of what was then an entirely new design capable of providing radio circuits having a traffic capacity greatly in excess of anything hitherto known in the sphere of long distance point communication. The stations built in the United Kingdom were at Bodmin, Dorchester and Grimsby,

The machinery room at the Bodmin Station, built by the Marconi Company for the British General Post Office. In the background may be seen the rectifying panels for the Canadian and South African transmitters.

all transmitting stations, with the receiving stations at Bridgwater and Skegness. Ongar, Brentwood and Somerton came along later.

Approximately 200 acres of land comprising the transmitting site were obtained and purchased from Viscount Falmouth by the Postmaster General, and work on the construction of the first HF Beam transmitting station in this country commenced in December 1924. The excavations, road building, fencing, foundations and construction of the buildings was carried out by the Foundation Company of London to the design of the Marconi Company Limited. Provision was made for the building of the Engineer-in-Charge's residence and a number of other dwellings, named 'Workman's Cottages' half a mile from the station.

The masts suspending the aerial arrays, developed by C. S. Franklin of the Marconi Company, were supplied and erected by the Armstrong Construction Co. There were five masts for each Dominion, in Bodmin's case five for the Canadian service and five for the South African service. The masts were 277 feet high, each having a cross arm at the top measuring 90 feet from end to end. The distance between the masts were 650 feet, and the length of the whole system of five masts was 3,200 feet. Basically a reproduction of the receiving aerial array at the Bridgwater station.

An early photograph of the Bodmin masts and aerials.

Power was self-generated and obtained from three Ruston and Hornsby cold starting diesel engines driving DC generators, each rated at 400 Volts, 200 Amps, situated in a room apart from the transmitting room. This room also contained alternators and HT rectifiers supplying power for the transmitters."

The station opened on schedule, at midnight on Sunday 24th October 1926. The Mayor of Bodmin, Mr. A. Browning-Lyne sent the following message to the Postmaster General, Sir W, Mitchell-Thomson:

"Mayor of Bodmin sends town's congratulations on successful opening of new Beam Wireless Station and hope it may be the forerunner of further developments linking all part of the Empire in still closer bonds of friendship."

View looking up at one of the Bodmin masts, c. 1926.

However, the opening of the Bodmin station did not go down well with many residents of the local area, when in December 1926, interference to broadcast reception was reported:

"Since the opening of the new Imperial Beam Wireless stations at Bodmin and Bridgwater last month, numerous complaints of interference have been made by West Country radio enthusiasts whose contact with the various broadcasting stations is practically rut off when the great "Beam" aerials are in operation. The interference is stated to be particularly bad in Cornwall, and Mr. G. Pilcher, M.P., has been bombarded with letters from Cornish 'victims of the beam' since he raised the matter in the House of Commons a few days ago.

Mr. Pilcher asked the Postmaster-General whether he had any information regarding the interference with ordinary wireless receiving in Cornwall by the new beam system of communication with Canada, and if he would approach the Marconi Company with a view to protecting the interests of the owners of receiving sets.

Sir William Mitchell-Thomson replied that the operation of the Post Office 'beam' station at Bodmin did not interfere with broadcast reception in Cornwell provided that suitable receiving apparatus was used. He agreed that some interference had been experienced by persons in Cornwall using short wavelengths for experimental purposes, but he thought that was unavoidable.

The Postmaster-General's reply has been strongly criticised by many Cornish listeners, who have written to

Mr. Pilcher describing the official defence us 'absolutely incorrect and misleading.'

One complainant, writing from St. Austell, states that he has a first-class set, and declares that interference takes place 'practically all day long on any wavelength. Daventry, Paris, Eiffel Tower, Dublin, and low-wave stations", he goes on, "are all the same. The noise is awful, though my set is perfect. Will the Postmaster-General explain how it is that on Sunday, November 14th, when the beam station closed down in the evening, our reception in Cornwall was perfect? We are all of the opinion down this way that it is the beam station at Bodmin that is causing all the trouble. Last winter we had no interference at all.

"I think," the letter concludes, "it is a shame that any company, municipality, or Government should be allowed to deprive people of a pleasure that they have spent a good deal of their hard cash in obtaining."

Another listener reports that the engineers of the B.B.C., who made investigations in Cornwall recently, were satisfied that there is interference which justifies the complaints of Cornish listeners. "Whatever the Postmaster-General may think, both the B.B.C. and the Marconi Company are aware of the fact that something is wrong, and they are acting accordingly. Of course, it is not possible for anyone to condemn the beam station, and naturally the officials do not like to think that this new station is capable of causing annoyance, hut there are numerous indications that this is the case."

The same correspondent has made several interesting tests, the results of which, he claims, prove conclusively that the interference is caused by the Bodmin station. 'V.L.' writing from Porth to the *'Western Morning News'*, refers to the interference from the 'amateurs using short wavelengths' mentioned by the Postmaster-General, in his reply to Mr. Pilcher.

In the first place, he says, the trouble on short-wave has now been overcome. Since the Post Office engineers took over the Bodmin station, the amateur bands have been clear of interference. In view of the intermittent interference on broadcast wavelengths, culminating in continuous interference on September 21, the matter was laid before the B.B.C. at a personal interview in London on September 26.

Further reports were obtained from many districts, and these were subsequently collated amid the results sent to Savoy Hill. In view of the serious trouble disclosed, an engineer from the Company was sent down to investigate on November 1st.

The trouble proved a most difficult one to locate, as it showed no directional effects and extended up to or over 20,000 metres with little variation in strength. There appeared possibly to be a slight peak between 1,000 and 1,100 metres.

In the meantime, through the help of Mr. A. M. Williams, the Postmaster-General has been approached, and with his sanction, and by the help of the engineer in charge of the Bodmin station, various tests were carried out. It should be mentioned that, from the first, there seemed to be some doubt as to Bodmin being the culprit. Several times it had been heard to close down when listening on 26 metres, whilst the interference on the Daventry programme, which was coming through on another set, in each ease was unaffected.

No connection between the period of the interference and Bodmin's marking of spacing wave could be located, although this point was observed on many occasions. There remained the possibility that either the absorption circuit or the alternators at Bodmin might cause the interference.

On the evening of November 12th, engineers from Bodmin visited us whilst the interference on Daventry was fairly bad, and phoned the station from this house, instructing that all power be cut off from both transmitters. This stoppage produced not the slightest effect upon the interference, thus proving the aerial and absorption circuits to be guiltless.

On November 14th and 21st, further tests were arranged to prove if other plant was at fault. On each day the entire station plant was stopped for roughly five minutes, shortly after 8 p.m. No difference was noticeable upon the broadcast wavelengths. In each of these tests, both the Canadian and South African beams have been observed from here direct on their own wavelengths to check times of closing down and restarting.

Further tests were arranged for the observation of the British Broadcasting Company's engineer on November 23rd, when traffic was stopped for four periods of five minutes each. It will be noticed that on the date mentioned by your St. Austell correspondent in a recent article (November 14), Bodmin was closed only for a few minutes. The general reception for the whole evening would therefore not be effected.

The interference, as shown by reports from many districts, has now greatly abated. The source, so far as we are concerned, remains unknown. Originally it was reported from Minehead to Pendeen, and from Torquay to Falmouth. It is not possible to anticipate the B.B.C.'s decision but, in view of the test results, of which they are cognisant, it seems the sources of the interference can scarcely have been at Bodmin. It must be remembered that there are other commercially-owned stations of an experimental character in Cornwall. Few will deny that the reception of Daventry has lately been bad beyond words. Many listeners will recall that Chelmsford also was afflicted with attacks of fading and distortion.

Radio Paris, with a fraction of Daventry's power, gives a better and consistent transmission. The latter station is well known to use much heavier modulation than Daventry, and get through much better with less power.

It appears that Daventry's carrier wave would stand at least 50 per cent greater modulation; at present its power is wasted to at least this extent. This view is borne out by the strike news bulletins, which were fully modulated, and came through at consistent strength, and without the fading and distortion since experienced.

Other 'special' programme items, such as the broadcast to America and Sir Harry Lauder's performances, can be cited. Although Swansea and Plymouth are now merged into the 'great howl', and Bournemouth and Dublin still further into the spark band, yet if the B.B.C. will run the Daventry station to give service to the distant parts of the country (for whom the station is surely intended), Cornwall would have a certain and powerful transmission fairly free from interference from other stations, and giving the pick of the English programme.. This would certainly be preferable to any number of relay or regional stations, liable to spark interference and with limited programme facilities.

In a letter on the same subject, Mr. Walter H. Graham, of Red House, Tywardreath, says:

"In fairness to the persons responsible for the working and adjustment of the Post Office Beam station near Bodmin, I think it should be made known that there are listeners to broadcasting in the neighbourhood of the Beam Station who have unable to detect any interference whatever on broadcasting wavelengths from the Beam Station. I have a high aerial, which collects most types of interference efficiently, situated about eight miles due south of the Beam Station and four miles east of St. Austell. I listen to broadcasting on most evenings at varying times.

Though the ways of the 'Beam' may be fantastic, one would expect. If the noise of it at St. Austell is awful, as one complaint would have us believe, there would be at least an echo of it here. I have spoken to experienced listeners at St. Austell and Newquay, who tell me they have detected no such interference, and Newquay listeners would be more likely to suffer than others being in the line of the Canadian Beam.

While not denying the possibility of interference in a queer manner, it is improbable that the Beam is responsible for the interference complained of. If your St. Austell correspondent who suffers practically all day long on any wavelength would communicate with me, I should be most interested to investigate, undertake comparative tests, and publish the results if he so desires."

A view of the aerial masts, Bodmin, c. 1926.

The South African service from Bodmin opened on the 5th July 1927 and the Beam Service was augmented shortly afterwards by the opening of services to India and Australia using the transmitting and receiving stations at Grimsby and Skegness.

One even which is unique in the history of the station and indeed in broadcasting, was the successful broadcasts of the Thanksgiving Service from Westminster Abbey for the recovery of King George V on 7th July 1929, by means of the Marconi-Mathieu multiplex system of telephony. The service was relayed to Bodmin by Post Office landline and beamed to Canada, then re-transmitted to Sydney where it was re-broadcast throughout Australia and to neighbouring countries by stations of Amalgamated Wireless Australia Ltd. (AWA).

The station came under the auspices of the Imperial and International Communications Ltd. on 29th September 1929, in common with all other Wireless Beam Stations formerly owned by the Post Office.

The stations at Bodmin and Bridgwater were used again to great effect in early 1930 when a broadcast by King George V from the opening of the Naval Conference was transmitted using the Beam Service. The occasion was marked by a detailed report in the Canadian press, which is an enjoyable read, the more so taking into account the objectivity and sense of wonderment from the other side of the Atlantic:

"As to sound, there is no more distance. The world is a small room in which people talk.

One of the masts and buildings at Bodmin, c. 1926.

On the third Tuesday of last January, His Majesty King George the Fifth drove to the House of Lords at Westminster, and there welcomed the delegates to the Naval Disarmament Conference.

Wireless Transmitting Station, Bodmin

An early postcard showing the Bodmin site.

As the King spoke, this reporter, seated before a simple piece of modest furniture in a Montreal apartment, listened to the words of his epochal address. Afterward, the voices of the Right Honourable Ramsay MacDonald and of the succeeding speakers were heard, with almost as perfect clarity as in a personal audience.

A million other Canadians, scattered from Cape Breton to Prince Rupert, from the Arctic Circle to the United States border, experienced the same amazing phenomenon at the same moment.

"Hello, Canada, here we are again"

Here we are, then, in front of our radio receivers on the morning of Tuesday, January 21, 1930. Because of the divisions of daylight and darkness imposed upon us by the earth's methodical journeyings around the sun, the clocks in Halifax show the hour of seven. In Montreal and Toronto it is six, in Winnipeg, five, in Calgary four, while on the Pacific Coast it is just three o'clock in the morning; but in London, Big Ben will shortly strike eleven, and by the alchemy of radio it becomes eleven o'clock the world over, since it is the moment when King George the Fifth will open the Naval Conference from which this war-torn and armaments-oppressed universe hopes so much.

As a preliminary to the big event, we hear first the strains of an organ recital, through which it is planned pleasantly to arouse eager listeners in the eastern half of the Dominion, while at the same time helping to keep awake those drowsy folks in the western half who have chosen to sit up late, defying sleep.

The music, we are told by someone, has its origin in the auditorium of a Montreal department store, but whether we are listening in at C.B. or B.C. over a trans-Canadian network of copper telephone wires, the recital is heard in Sydney and in Vancouver alike as clearly as though broadcast from the nearest local station in its original volume and tone.

At intervals a voice, invisible, disembodied, but distinct, as it might be from the next room, remarks, "Hello, Canada. Here we are again," and that is the key announcer, W. D. Simpson, warning us to be alert, because "we will give you London very shortly now." The announcer from our local station also breaks in between numbers to let us know that our interests are being protected by the mystic combination of call letters with which we are day by day familiar.

At last; "Stand by for London. Coming in at once." And then the voice of the London master of ceremonies remarking conversationally, as though he were telling that it looked like a fine day along the Thames Embankment:

"We will now take you over to the gallery of the House of Lords."

Finally, the actual voice of King George the Fifth, speaking to hundreds of thousands of loyal Canadians, who, but for radio, might have lived their lives and found their graves without hearing his voice.

Not only that. Throughout Europe, in the Antipodes, in each of the forty-eight states of the American Union, in South America, South Africa, India, Japan, China - the world around the King's address and the subsequent proceedings were heard in this first universal radio broadcast.

This a miracle. It is also a fact of common knowledge, as certain as sunshine or darkness. A new world, in which distance no longer exists. The superlative genius of Morse, Bell, Edison, Marconi, and de Forest combined, supported by the unobtrusive researches and unremitting labours of hundreds of other scientists whose names are less popularly known, has squeezed the terrestrial globe into this insignificant compass since that day in 1876 when Alexander Graham Bell obtained the first telephone patent, and so established the practicability of instantaneous transmission of the human voice through the medium of electricity.

We call it radio; but radio means the wireless projection and reception of sound through the air. That is by no means all there is to radio in its modern development. The world-wide radio broadcast of 1930 is more than fifty per cent a matter of sound transmission over delicately attuned copper wires, as this astonished reporter quickly discovered when a few weeks ago, on behalf of MacLean's Magazine, he set out to discover by what legerdemain the King's broadcast was made possible.

Two distinct and different systems were used to speed the addresses which opened the Naval Conference across the world's oceans from England - the shortwave radio and the Marconi beam transmission. In Canada most listeners received their broadcasts over the Marconi beam. This was the case also with other parts of the Empire. Between England and the United States the broadcast was made by short wave. In both cases land transmission from the receiving points to local broadcast stations was accomplished through a maze of telephone wires, controlled by various telephone and telegraph corporations.

Our Canadian broadcast involved the co-operation of one department of the British Government and five commercial companies: the British Post Office, controlling telegraph and telephone services in the United Kingdom: the British Broadcasting Corporation, holding a monopoly, under governmental authority, of radio broadcasts throughout the British Isles; the Marconi Wireless Telegraph Company of London; the Canadian Marconi Company; the Bell Telephone Company of Canada, and the Canadian National Telegraphs. Exact co-ordination of the services of each of these organizations was absolutely essential to the success of the project. Canadian listeners from coast to coast are able to testify with eloquence that the difficult task was magnificently accomplished.

Thousands of miles of copper wire, accurately linked with 3,000 miles of Marconi beam transmission across the Atlantic, made the Canadian broadcast possible. The King's speech and those of the delegates travelled to this country, first by telephone wire from the House of Lords in London to Bodmin in Cornwall; then by Marconi wireless beam to Yamachiche, Quebec; then by telephone wire to Montreal, through which city the entire Canadian broadcast was relayed to twenty-one local stations, from CJCB, which is in Sydney, N.S., to CNRV, which is in Vancouver, B.C., over a network of wires of the Bell Telephone Company of Canada and the Canadian National Telegraphs. Each receiving station handled its own broadcast. Reception everywhere was, for all practical purposes, instantaneous not only in Canada, but the world over.

Attempting to understand this incredible thing, the non-scientific mind has first to discard all previously accepted ideas of what constitutes speed in relation to distance, when sound is under consideration. Simple folks, such as myself, hearing that Major Sir Henry Segrave has driven an automobile at more than 230 miles an hour, find the information wonderful but not beyond belief, because we have ourselves driven or ridden in automobiles travelling at fifty or sixty miles an hour. When we hear that aviators of the British Schneider Cup team have hurled an airplane through space at more than 330 miles an hour, that is marvellous, but we do not doubt it, because we know that every day Canadian aviators are driving their machines at 100 miles an hour or over.

But in this case we have to forget miles or hours, or even seconds. The King spoke in London and Bombay and Winnipeg, Buenos Aires and New York, San Francisco and Melbourne, Rome and Cape Town, Tokyo and Copenhagen, and Montreal and Madrid heard his voice on the instant.

There is no more space. Radio and the telephone, united, have destroyed distance.

My long-suffering mentors, L. S. Payne, of the Canadian Marconi Company, and LeSueur Brodie, of the Bell Telephone Company of Canada, assure me that it is possible that the King's words were heard by Canadian listeners a fraction of a second before they reached the ears of those members of his audience who chanced to be seated farthest from the dais upon which he stood in the Royal Gallery of the House of Lords.

To the trained mind this is rudimentary; but the man on the street finds it difficult to assimilate. The fact is that the electrical waves, transmitted by radio or over telephone wires travel at the speed of light, which is, approximately, 186,000 miles per second. The ordinary speaking voice, with a range of but 1,100 feet per second, is a snail compared with this; which primary fact makes it possible to speak in Halifax over the telephone, and be heard instantly in Vancouver. In effect, the broadcast of the Naval Conference's opening was the longest long-distance telephone talk in history.

Another important factor to be borne in mind is the circumstances that to these radio people, most of them young men who appear to regard their job of working miracles as casual routine - no more exciting than keeping books or selling teakettles - any broadcast programme is a tangible thing, having form and substance. They carve it up as father carves the joint, and serve it around the country to Saskatoon or Sydney, Charlottetown or Calgary, as simply as mother divides the apple pie among a hungry family. A regiment of them carved and apportioned the King's broadcast in exactly this fashion. Other regiments passed the same programme around the world.

His Majesty spoke before a microphone, as, of course, did the eminent statesmen of the five great powers and the British Dominions who followed him. The microphone carried the sound waves first to a near-by amplifier, then, in the case of the Canadian broadcast, to telephone wires, which conveyed them to GBK, the Marconi Short Wave Wireless Beam Transmitting Station for Canada, which is located four and a half miles outside the ancient town of Bodmin, in Cornwall. Bodmin, the county seat of Cornwall, dates back to the Roman conquest. Had any of those old empire builders been present in the flesh on January 21, what a dumbfounded lot of Romans they would have been!

From the towering latticed steel transmission masts of Bodmin, the Marconi beam shot the message into the air, directing it toward the little French-Canadian village of Yamachiche, located on the north shore of the St. Lawrence, some thirty miles east of Montreal. At Yamachiche is established the Canadian receiving station for the England-to-Canada beam service - the target toward which the Bodmin beam is directed. The station call letters are CGA.

CGA at Yamachiche took the broadcast from Bodmin, transferred it to telephone wires, and so passed it to Montreal, to the downtown Bell Telephone office on Notre Dame Street.

Here Horace Belanger, wire chief, and his assistant, Joseph Desgroseilliers, took hold of the message, with W. D. Simpson doing the announcing. Mr. Simpson is manager and chief announcer for CFCF, the Marconi broadcasting station on the roof of the Mount Royal Hotel, but for this occasion he had moved himself and his microphone downtown to the Bell Building, because this is where the programme arrives from Yamachiche to be redistributed across Canada.

Here the incoming sound waves were split twice, once for the local Montreal broadcast through CFCF, and once for the head office of the Canadian National Telegraphs on St. Sacrament Street, just around the corner. The split was handled through a broadcast repeater panel of the latest type, a ten foot-high nest of wires, dials and switch plugs. All the operators occupied with this process were constantly in immediate touch with each other through direct line telegraph and telephone wires. If it was inconvenient to talk, they used Morse signals; if the Morse was not available they talked.

At Canadian National Telegraphs headquarters, the heart of the trans-Canada network in this instance, operator "Andy" Anderson listened, modulated and routed the programme east and west. He was also in touch by telephone with the Bell broadcast repeater panel, and by phone and telegraph with Ottawa, Toronto and Quebec. His board took the sound waves from the telephone company's lines and split them three ways, sending to Ottawa, Toronto and Quebec simultaneously.

From Toronto westward, with a local broadcast through hooked-up stations at each relay point, the programme was sent to London, Hamilton and Winnipeg, through Winnipeg to Yorkton, Regina and Saskatoon; through Saskatoon to Edmonton and Calgary, and so to Red Deer and Vancouver. Eastward from Montreal, the message travelled to Quebec and from Quebec to Moncton, to Halifax and Sydney. In each case, between main stations, the sound waves were carried over balanced telephone wires, which differ from telegraph wires in that they are capable of transmitting music and the human voice as well as Morse signals. Telephone and telegraph companies alike use both types of wire.

Certainly such precise coordination as was required to achieve this broadcast establishes a new peak in human accomplishment, more than ever when multiplied by the sum of the transmission and retransmission organizations involved in the broadcast to the rest of the world.

But the boys have other conjuring tricks in their repertoire, should emergency arise. It so chanced that the operators concerned in the Yamachiche and Montreal relay were required on this extra special occasion to perform one of the most amazing quick-change acts in the brief history of wireless communication.

Canada is linked with Australia over the Marconi beam, as well as with Great Britain. The transmission station which connects this country with the Antipodes is CJA, located at Drummondville, Quebec, on the south shore of the St. Lawrence, thirty miles east of Montreal and twenty-five miles south of Yamachiche. The station is equipped to handle both telephone and telegraph messages over the beam.

England is in direct communication with Australia through a transmitting station established six miles outside Grimsby, in Lincolnshire.

Weeks before the final plans for the Naval Conference beam transmission throughout the Empire were announced, Marconi engineers considered the advisability of routing the service to Australia by way of Canada through Drummondville. The decision was to use the direct route from Grimsby.

Something went wrong, and a few minutes after the Grimsby transmission began, a message from Australia reported poor reception. An immediate decision to re-route the programme through the Drummondville station followed.

Read this paragraph slowly: Grimsby communicated by wire with Bodmin. Bodmin advised Yamachiche by beam that the Grimsby transmission to Sydney was giving trouble, and that Drummondville must take over. Yamachiche relayed this message by wire to Montreal. Montreal passed the word by wire to Drummondville.

Drummondville reported to Montreal by wire, "All ready to take over." Montreal advised Yamachiche by wire. Yamachiche reported to Bodmin by beam. Then at a given signal Grimsby stopped sending, Drummondville took the programme by wire from Montreal, and continued to send the balance of the broadcast to Australia until the "sign off" signal was received.

The total elapsed time between receipt of the first message at Yamachiche and the establishment of the Bodmin-Yamachiche-Montreal-Drummondville-Sydney service was ten minutes. And to think that in days gone by we stood popeyed, as a stage magician pulled an ordinary rabbit kicking from an ordinary silk hat!

The Canadian service suffered no interruption or interference. The only inconvenience was inflicted upon a squad of faithful operators and engineers at Drummondville, who were dragged from well-earned slumbers and had to make the necessary switching operations in their pyjamas.

This sort, of round the world hopscotch is, to my ingenuous mind at least, not less than fascinating. We know it was done - but how?

Thanks to the remarkable patience of a number of sound engineers with my feeble attempts to understand their wizardry, this article will attempt to explain in the simplest terms passible the operations of the Marconi beam system of wireless transmission, which is, the experts declare, the most important development since wireless was first demonstrated as practical. The Marconi beam is of the utmost importance, not only to Canada, but to the whole of the British Empire, since it is more generally used for transmission between the various British nations than elsewhere. You can pick up an ordinary telephone instrument in the basement of the Marconi Building in Montreal today and talk immediately to London. During the Naval Conference, Colonel J. L. Ralston, the Canadian delegate, was in daily communication with the Prime Minister at Ottawa over the beam. For reasons not pertinent to this article the system has not yet been generally commercialized; but it is established, just as surely as Canadian listeners heard King George talk over the beam last January.

The feature of the Marconi beam which differentiates it from the short-wave method, is that the beam is directed toward a definite objective, while the short wave is a broadcast. To the lay mind this may be shown as the difference between the distribution of light rays from a standard ceiling electric lamp, and from a focused searchlight. In the first case, the light rays are spread in all directions. In the second, they are concentrated in a narrow band and aimed at a target. A broadcast is in wireless what a standard light fixture is in electric lighting. The beam is the searchlight, and from this similarity it receives its name.

The effect of focused wireless waves is achieved by the use of a curtain of closely placed wires behind the sending and receiving installations. These act in the same capacity as the mirror or polished metal reflector behind the searchlight, and Marconi engineers, again borrowing a familiar term in lighting engineering, describe them as reflectors.

The use of reflectors to increase the range of wireless stations was considered by Senatore Marconi as far back as 1895, but lacking the present day knowledge resulting from subsequent exhaustive research and experimentation, and with commercial considerations also entering, reflector experiments were pushed aside in favour of the development of high-power long-wave stations.

During the war, the British government eagerly sought a means of wireless communication which would make the interception of messages by hostile agencies impossible. Senatore Marconi, with C. S. Franklin collaborating, resumed reflector experiments, and successfully established the practicability of the beam idea.

In 1923 beam communication was established between the Marconi station at Poldhu, in Cornwall, England, and Senatore Marconi's yacht *'Elettra'* far out at sea. Later, long-distance tests were successfully carried out between Poldhu and Sydney (Australia), Buenos Aires, Rio de Janeiro and Drummondville. Canadians have every reason to be proud of the prominent part which the Dominion and the Canadian Marconi Company have played in the present amazing development of wireless telegraphy and telephony. The Drummondville and Yamachiche stations are vital links in a chain of instantaneous intra-Empire communications of supreme importance.

The Marconi beam, sound engineering experts assure me, is, right now, by far the fastest method of communication yet devised. Its speed is limited only by the mechanical limitations of manipulating and recording instruments at each station. Restriction of radiation to a narrow beam, the screening effect of the reflector at the receiving station, and the large number of wave bands available make it possible to take care of a great number of separate services.

It is possible to send over the beam, and to receive, telephone and telegraph messages at the same time without interference, and the beam can be used, and is being used, for facsimile transmission over any distance.

Very soon now, it will be as simple a matter to call up London, England, from London, Ontario, as it is to telephone Ottawa from Toronto. Photographs, documents, even fingerprints may be transmitted by wireless.

This is going to be a very uncomfortable world for the international crook.

Long-distance radio transmission is not, of course, confined to the Marconi beam system. The short-wave broadcast, used by a large number of stations throughout the world, has kept step with the general advance of radio science, and has achieved, and is still achieving some astonishing results. The Naval Conference broadcast to the United States and to other foreign countries outside the British Commonwealth was a short wave omni-directional proposition, co-ordinated, as in the case of the beam, with telephone and telegraph land transmission.

The great majority of radio receivers now sold commercially are long-wave sets. The general public, which regards radio rather as entertainment than as an absorbing field for experimentation, is almost entirely unaware of the remarkable achievements in distance short-wave broadcasting recorded by Canadian stations and by some of our amateur radio experts, who have either constructed their own short-wave receiving and sending sets, or have installed short-wave adapters on their long-wave machines. One enthusiast residing at St. Lambert, Quebec, was in frequent communication with Commander Byrd's party in the Antarctic during the winter months.

The average broadcast wave, which brings entertainment from comparatively nearby stations to your loud speaker, is about 300 metres long. A metre is slightly longer than a yard. Short-wave broadcasting means the use of waves which are between twenty and eighty yards long.

By this method, short-wave stations can be heard in the immediate vicinity up to a distance of approximately sixty miles. Then comes a skip of from 400 to 600 miles over which the short waves fail to register; but from that radius on there is practically no limit to the reception range. The usual broadcast over the long-wave system can be heard from next door to the transmitting station up to a range of about 1,000 miles or a little beyond, depending upon several contributory circumstances, the strength of the sending station, the power of the receiver, and reception conditions in the atmosphere among them.

Station CFCA, owned and operated by the Toronto Star, inaugurated a policy of picking up short-wave transmission from overseas for re-broadcast in the Toronto district, early in 1928, and has successfully carried on with the re-transmission of English programmes daily since then.

One of the most famous of Canadian short-wave stations is CRJX, located at Middlechurch, Manitoba.

This is one of a chain of three stations operated by James Richardson and Sons, of Winnipeg, radio pioneers in the West. The others on this chain are CRJM at Moose Jaw, Saskatchewan, and CJRW at Fleming, Saskatchewan, both long wave installations.

The log of CJRX reports reception of its programmes, broadcast from Winnipeg, in Great Britain, South Africa, Australasia, the West Indies, several South American countries, and the Belgian Congo. That gives the average radio listener something to remember the next time he is tempted to boast to his friends that he got Cincinnati last night.

This article dealing entirely with long distance radio has inevitably to discard a wealth of interesting material about the details of studio management and equipment, hook-ups, and instances of the unusual uses to which radio in its present form has been put. Perhaps there will be an opportunity later for this. The thing which persists in this reporter's mind as paramount, is the enormous strides which radio has made in the last five years. In point of perfection of mechanical equipment, simplicity of operation yes, and quality of service offered to the general public-radio has progressed more rapidly in five years than the automobile did in twenty. Such miracles as the King's broadcast and its reception in Canada prove this to be true.

Nor is the end yet in sight, for television is just around the corner, radio engineers say. After studying the map of their achievements in the last decade, this ignoramus, at least, is quite ready to believe any statement they care to make."

Services continued successfully under the new Company (now known as Cable and Wireless) until the outbreak of war in 1939 when it was closed and completely vacated by the owners. The transmitter room and engine house were completely stripped and the transmitters were ultimately relocated at the Dorchester transmitting site. Only the Beam masts and aerial arrays were left.

In 1940, the station was leased to the Air Ministry and the Royal Air Force occupied the premises, installing their own HF radio equipment and aerials, and continued to operate the station until cessation of hostilities in 1945.

In 1947, Cable and Wireless were asked to re-open the Bodmin station by the British Admiralty, and to install transmitters provided by them and maintain services operated from Whitehall to Naval ships and bases overseas.

Under the new fitting, a total of twelve transmitters were installed, including two low-power transmitters for coastal working, remotely-controlled from Plymouth Marine Headquarters. The aerials, a mixture of dipoles and rhombics of Naval design, were supported by 150 foot Marconi steel lattice masts and wooden towers.

Telegraph services were operated to naval bases at Malta, Gibraltar, Bombay, Karachi, Mauritius, Canada, Australia, and the NATO base at Northwood, Virginia, USA as well as ships broadcasts.

At the end of 1949, under the 'Nationalisation Scheme' introduced by the Government, Bodmin Radio returned to Post Office ownership, which brought about the precarious task of integrating Post Office staff with ex-Cable and Wireless staff, as well as sorting out grievances and anomalies that this brought. Hours of attendance, annual leave and rates of pay had to be integrated, similar to what occurred under the famous Act of 1929.

A few months before Nationalisation, Cable and Wireless had placed a contract with the Marconi Company for the supply of new transmitters. These were eventually installed by station staff and became operational in 1953, providing Post Office telegraph and facsimile services to South Africa and Canada using the original Franklin aerial arrays.

In May 1961 these aerials became redundant and the 277 foot masts were taken down and replaced by Post Office aerials of rhombic design. In the same year, the main transmitting room was extended to accommodate a further four 4 kW Naval transmitters.

New smaller aerials were added in 1967, and at the end of the decade, the Ministry of Defence (Naval) decided upon a replacement transmitter programme for the station. The existing transmitters were withdrawn and replaced by modern 30 kW units and back-up equipment. The transmitters selected were Marconi H1200 units, already in use at Naval Shore wireless stations at home and abroad. These were to be installed by the Post Office but due to staff shortages and other work, the installation was sub-contracted to the Radio Corporation of America (GB) Ltd., better known as RCA.

General view of the Bodmin transmitting station.

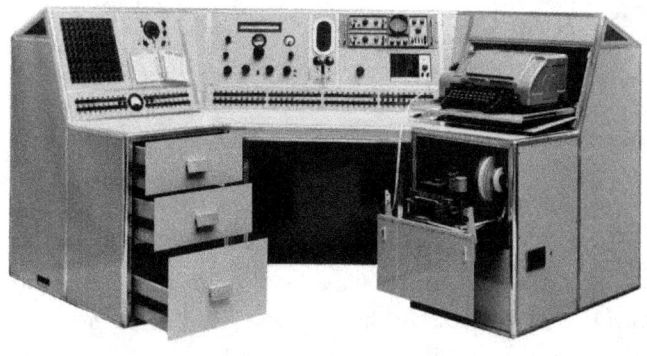

1964 monitoring control console, Bodmin Radio Station.

breakthrough

MST 30kW transmitter type H1200

An h.f linear amplifier transmitter for high-grade telecommunications.
Frequency range; 4-27.5 Mc/s.
Output power: 30 kW p.e.p, 20 kW c.w,
Meets all CCIR Recommendations.

saves 80% floor space

Transmitters can be mounted side by side and back to back or against a wall. Floor-ducts are eliminated and all power supply components are built-in. These features lead to smaller, simpler, cheaper buildings or more services in existing buildings.

rugged reliability

R.F circuits have been simplified and the number of mechanical parts reduced to a minimum. Highest engineering standards are applied to the design of these parts; stainless steel shafts in ball-bearings in heavy, rigid, machined castings; stainless steel spur gears meshing with silicon bronze; heavy r.f coil contacts with high contact pressure. Specified performance is maintained with ample margins.

simplicity

MST reliability allows continuous unattended operation with extended or remote control, saving maintenance and operating staff. Any fault in the servo control circuits can quickly be located with simple test routines. Transistors and printed wiring give these circuits maximum reliability.

self-tuning

The H1200 has a frequency following servo tuning system. Any frequency may be selected on the synthesizer decade dials in the associated MST drive equipment; the unattended transmitter automatically tunes itself in an average time of twenty seconds. Final stage tuning and loading servos continuously ensure automatic compensation for changes in aerial feeder impedance caused by weather conditions. Self-tuning gives one-man control of an entire transmitting station.

Advertising material for the Marconi H1200 transmitter.

The last transmitter was installed in June 1970, but the MoD gave notice that the station would close in 1973. On receipt of this information, the Post Office gave permission to extend the A30 road from the terminal of the new Bodmin by-pass, by constructing a road through the station site. When correspondence on this matter was re-opened in 1973, the MoD had a change of heart and advised that closure of the station would now not take place and the planned A30 extension was unacceptable. At a meeting in Bodmin on 3rd April 1973, the MoD stated that the location of the station was extremely advantageous and if it was to close, an alternative site would need to be found in Central Cornwall. Use of the station for MoD purposes remains shrouded in secrecy but it has been reported that the station carried a 'hot-line' between the UK Government and Moscow until this link was transferred to secure satellite links, but official confirmation of this is difficult to obtain.

1988 saw the arrival of small dish satellite aerials, no doubt installed as part of the ongoing MoD requirements.

The station continued in use for the Post Office (later British Telecom International and thence BT) providing long-distance radio links until 2002 when, overtaken by technology and satellite communications, it closed on 31st March 2002. The aerials remained for a short time but were finally removed around 2005. The building has been derelict since closure and was sadly subject to a significant amount of vandalism, which put at risk a number of protested species.

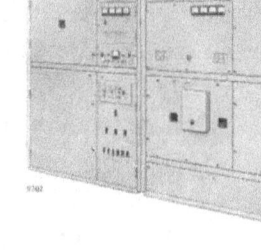

THE H 1200 is a 30 kW p.e.p linear-amplifier transmitter having a self-tuned final stage and untuned, wide-band amplifiers for the carrier stages. When the drive frequency is changed, the self-tuning system operates to provide the correct tuning and loading within about 20 seconds. The output tuning and loading servos remain in operation at all times to ensure accurate aerial matching in spite of changes in aerial impedance as a result of varying weather conditions.

Only three variable tuning controls are used for matching the input to the final stage and tuning and loading the output circuit, thereby facilitating the adoption of fully automatic self-tuning techniques. In addition, the input level to the amplifier is automatically regulated. Maximum reliability is ensured by keeping the number of moving parts to a minimum.

The controls are operated by high-quality transistorized servo systems which derive their inputs from tuning and loading error detector circuits in the amplifier.

The power supplies incorporate the latest techniques in silicon rectifiers and air-cooled transformer design, enabling all power components to be incorporated within the transmitter and avoiding the necessity for special accommodation.

Features

Self-contained, no back or side access required.

All supplies use compact silicon rectifier assemblies.

Full feeder protection and output maintenance, with built-in directional coupler.

Full extended control facilities available.

Data Summary

Frequency range: 4–27·5 Mc/s.

Services: i.s.b, s.s.b, f.s.k, f.s diplex, c.w on/off (depending on type of drive).

Output power: 30 kW p.e.p ($\pm$ 0·5 dB) on i.s.b, 20 kW ($\pm$ 0·5 dB) on c.w on/off and f.s telegraphy without manual adjustments.

Harmonic radiation: No harmonic emission exceeds 50 mW. A low-pass filter fitted in the output feeder reduces the harmonics above 30 Mc/s.

Output impedance: 50 Ω unbalanced, max. v.s.w.r 2:1.

Noise level: (a) More than 30 dB below carrier level (carrier normally −26 dB to −16 dB relative to p.e.p) for components up to 200 c/s either side of carrier.
(b) More than 50 dB below p.e.p for all components of a single tone up to −6 dB relative to p.e.p.

Non-linear distortion: All (p's better than −36 dB relative to either of two equal tones for any power level up to full p.e.p.

Pilot carrier compression: Less than 1 dB for any level of single frequency signal up to −6 dB relative to p.e.p.

Input from drive: 25 mW minimum, 2 W maximum in 75 Ω at radiated frequency.

Suitable drives: H 1600 series (page 230).

Power supply: 380/440 V, 3-phase, 4-wire, 50 or 60 c/s. (as ordered).

Power supply variation: Except where station supplies are regulated to within $\pm 1\%$, the transmitter is supplied with an automatic voltage regulator which is normally set to accept mains variations of −6 to −14%.

Power consumption: Mark 65 kVA, space 26 kVA, i.s.b. 55 kVA at 0·93 power factor.

Climatic conditions: 45°C dry heat, 40°C at up to 90% humidity, maximum altitude 6000 ft. (1830 m).

Dimensions:

Height	Width	Depth
7 ft.	12 ft.	2 ft. 6 in.
(213 cm)	(366 cm)	(76 cm)

The Marconi Company Limited
Marconi House, Chelmsford, Essex
Telephone: Chelmsford 3221 · Telex: 1051
Telegrams: Expanse Chelmsford Telex

Technical details of the H1200 transmitter.

It was reported in September 2019 that the station buildings were to be restored and given a new lease of life:

"There are plans to revitalise a derelict radio station in Cornwall which was once illegally used for a rave party. The Bodmin Moor Radio Station, by the A30 near Lanivet, will be given a second life after new plans have been approved by Cornwall Council.

Carter Jonas, the national property consultancy, has secured planning permission on behalf of GAP Group Limited (GAP), the national plant and tool hire business. The proposals safeguard the future of the site's buildings by refurbishing the historic transmitter hall and canteen block."

They also retain an ecological buffer zone around the edge of the site and provide landscape planting, a bat tower, reptile refugia, and interpretation boards to promote understanding and appreciation of the site's history.

The company did indeed take over the buildings in 2020, and continue to use the site to this day.

Bodmin Radio Station, 2019.

CHAPTER 8

SOMERTON WIRELESS STATION

The first mention of a wireless receiving station in Somerton came in November 1925 when Mr. R. W. Pretor-Pinney gave a speech to members of the Somerton Cricket Club:

"Messrs. Marconi have bought 150 acres of land from me for the erection of a wireless station. I signed the conveyance today, so I do know it to be correct. (Cheers).

The land has been pegged out for the setting up of a wireless station, one serial to receive messages from North and the other from South America. The aerials will each be supported by five masts made of steel, each of which will be 277 feet high. On the top will be an arm 90 feet long from which wires will run to the power station, and from that lower station messages will be sent to Radio House, London, and from there sent to their destination in different parts of the country. That will bring a good deal of trade into Somerton. (Cheers).

I am told it will need about 100 men to put up the plant, and in all probability the permanent staff will be twelve people, because it will be a night and day station. Those people will want houses, and I think Mr. Spink or someone else who puts up the houses will do very well. That station will be sure to bring up some people to Somerton to have a look at these masts. I do not think we shall ever on account of its situation, make Somerton in industrial centre, but if we can't make Somerton an industrial centre let us make it a residential one. (Hear, hear).

If we have more people living here it will circulate more money, and people will become prosperous."

Although not too far from the Bridgwater station, Somerton's aerials were designed for reception from the Americas, which the Bridgwater versions would not have been suitable. Similarly, Bridgwater's sister transmitting station at Bodmin would have had the same issues, resulting in a transmitting station at Dorchester being used for the Somerton service.

Further details came in the local press from April 1927:

"A Beam Wireless station is being erected at Somerton, a former county town of Somerset, for the reception of wireless messages from New York and Rio de Janeiro. The site of approximately 165 acres was acquired from Lord Ilchester and Mr. R. W. Pretor-Pinney, and is north of the Langport and Yeovil main road.

The installation will be a double receiving station acting on the beam, 'direct wireless' principle. This double aerial system will consist of two lines, each of five latticed steel masts 277 feet high, with cross arms at the tops 90 feet long, and the length of each aerial will be about half a mile.

Near the junction of the mast lines, practically forming an obtuse angle, the cabin of the station will be erected.

The building, which will be in two parts, connected by a short corridor, will contain the receiving apparatus for the two lines of aerials, with a small charging plant driven by an internal combustion engine. The building will be connected with London by telegraph lines. Messages received on the aerials from North and South America will transmitted to London. The Somerton Beam Wireless station, which will be solely a receiving station, will be situated about halfway between the wireless stations at Dorchester and Chedzoy (near Bridgwater)."

A progress report appeared locally on 24th December 1927:

"The new Marconi receiving station at Somerton consists of a large building which will contain all the receivers of the beam services 'Via Marconi' and the necessary landline gear. The lines of masts are similar to those at the Dorchester transmitting station, except that no masts for the reception of signals from Japan will be erected. The signals from Japan will be received on aerials suspended on one side of the masts used for the reception from South America and the reflector between the two aerials will be common to both. The reception from North and South America at present takes place in a temporary hut containing the receivers, which are connected by landlines to Radio House. The wavelengths which will be used by the Marconi Beam Services will be between 13 and 40 metres."

An early overview of the South American Beam Wireless service with special emphasis on the Somerton receiving station was published in the February 1928 issue of 'Modern Wireless' and is well worth reproducing here:

"The South American Beam Services are entirely new services, and place Brazil and the Argentine in direct wireless communication with London for the first time. Having regard to the enormous amount of business conducted between London and South America, the opening of this direct high-speed beam service is of the greatest importance to business organisations in all the countries concerned. The beam service operates both to and from Rio de Janeiro, but at present Buenos Aires has no beam aerials, so that while messages are sent from London by beam to Buenos Aires, there is no beam in the reverse direction. It is expected, however, that this service will be made into a complete two-way beam service in the near future.

A good deal has been said recently with regard to the astonishing success of the Beam Services, which, on the authority of the Australian Prime Minister, are said to have attracted 45 per cent of the Pacific Cable Co.'s traffic from Australia alone, apart from the amount it has attracted from other cable routes: The immense strides that have been made and the large and increasing volume of traffic that is being carried by the beam will be realised from the fact that, during the week ending December 3rd, the total number of words carried over the four Empire circuits was at the rate of considerably more than 30,000,000 words a year. These services are operated from Radio House by remote control, which means that the wireless transmitters are controlled from the signalling keys in London. The beam transmitting stations for the North and South American services are situated at Dorchester, and the receiving station at Somerton, about 30 miles from Dorchester. By means of a simple relay at the transmitting station the signals sent from London over the connecting landline operate the wireless transmitter from which the signals are finally despatched through space to the distant receiving stations.

Similarly, the signals received on the beam in receiving aerials at Somerton are instantaneously conveyed by landline to Radio House, where they are recorded on paper tape in the form of Morse signals, and then typed on to a telegraph form for delivery. These services are normally worked at between 100 and 200 words a minute, although much greater speeds are attainable and over 300 words a minute have been worked quite satisfactorily during tests. Over the wireless circuits alone - when signals have not to be relayed over landlines - almost unlimited speeds are attainable. The only limit appears to be that introduced by the recording apparatus that is used.

It is possible by electrical and engineering adjustments to vary the width of the wireless beam, and the beam transmitting aerials that have been built by the Marconi Company at Rio de Janeiro for communication with Europe have been so constructed that London, Paris, and Berlin are brought within the span of the beam when it reaches Europe. In this way the Rio stations can handle traffic for all these three capitals. Beam receiving aerials are now being built near Paris for receiving from America, but none are yet in existence in Germany. It is not possible at present, therefore, to work with these capitals at as high a speed as can be worked with England, where the beam receiving aerials are in use.

Beam stations are also being constructed at Dorchester for communication with Egypt, in Egypt for communication with England, and at Dorchester for communication with

Japan and the Far East. The Dorchester and Somerton stations are thus centres from which numerous services will be conducted. When the final arrangements are completed there will be seven Marconi Beam Services operated from these centres: one to Egypt, two to the United States of America, two to South America, and two to Japan and the Far East. At Dorchester there is a row of five masts, so arranged that the great circle bearing on the North American and Egyptian stations is at right angles to the line of masts. Three bays on one side of the line of masts are used for transmission to North America and two bays on the other side will be used for working to Egypt, the Egyptian transmitter being arranged for working on two alternative wavelengths.

The two active aerials for the Egyptian service and two of the active aerials for the North American service are built with a reflector common to both and placed between them. There are two other masts, which are erected in a line at right-angles to the great circle to South America, and these carry the aerials and reflectors for the transmission of signals to Rio de Janeiro and the Argentine. Two more masts are in the course of erection, and will be used to suspend the aerials and reflectors for transmission to Japan."

The operation of the station necessitated the installation of new landlines, and in February 1928 it was reported that: "A considerable amount of work has been carried out on the land lines between Somerton Wireless Station and the Wireless Station at Dorchester, and between Somerton and London."

The station was proud to be involved in the local community, and in April 1928, about 40 members of the Yeovil School Science Society met at the station. The visited was recorded locally as a report of the time recalled:

"The visit was preceded by a lecture, the Marconi Wireless Telegraph Company lending an excellent series of lantern slides, and with them detailed information regarding Beam Wireless and the Marconi stations. The party, by permission of Mr. H. M. Burrows, engineer-in-charge, were taken over the whole station, from the aerials and reflectors where signals are picked up, to the tapes at the automatic inkers, where the signals are watched and checked.

The station is still working on temporary apparatus, but the party was shown over the power house, and the permanent receiving rooms, which are in an advanced state of construction. At the time of the visit signals were being received from beam stations at New York and Rio de Janeiro, but the station is being equipped to deal with messages from Japan, Cairo and Buenos Aires, in addition to the present two. The thanks of the Society were expressed to the Marconi Company for permission to pay the visit, and for the kindness shown in providing facilities for the lecture; to Mr. H. M. Burrows for his help in preparation for the lecture and to him and his two assistants who conducted the party round the station."

Trials of facsimile transmission were undertaken at the Somerton site in 1929, as reported in the May 1929 issue of *'Wireless Constructor'*:

"The wonders of the new Marconi system of facsimile picture transmission were demonstrated recently by Marconi engineers in a tiny hut at Somerton, Somerset. This new development, by which it is hoped ultimately to be able to transmit telegrams in the actual handwriting of the sender, should undoubtedly be regarded as a milestone in the successful transmission of pictures by Beam radio. For three years the research department of the Marconi's Wireless Telegraph Co., Ltd., has been experimenting with a view to producing a scheme by which it would be possible to commercialise picture transmission by radio on a larger footing than at present, and from the results that we were able to see it would appear that the day of 'Facsimilegrams' is not far distant.

There are at present certain technical difficulties in the matter of finding suitable landlines for relaying the pictures from the Beam receiving station to various parts of the country, but as far as the wireless link is concerned, a stage has now been reached when it is possible to send a picture 10 in. by 8 in. across the Atlantic in ten minutes or less. The pictures received at Somerton during the course of the demonstration included cartoons, fashion plates, sections of newsprints, etc., and in every case the quality of the received image was of a very high order. It is interesting to record that a considerable amount of trouble has been taken by the Marconi Co. in their experiments to determine the wavelengths best suited for transmission at different times of the day, and that it is sometimes necessary to change the wavelength as many as three times during the twenty-four hours in order to obtain a constant service."

The actual method of transmission and reception was extremely complex for its time, but a concise explanation of the methodology used was found in a brief article found in the February 1929 issue of the *'Marconi Review'*:

"For the purpose of establishing the service the picture transmitting equipment has been installed at the Canadian end of the beam circuit and works to the beam receiver located at Somerton, in Somerset.

To those thoroughly acquainted with picture transmitting systems the Marconi apparatus may be briefly described. Picture analysis is effected by a rotating and traversing spot of light. Modulated signals are created with an interrupting carrier frequency and photoelectric cell. Synchronising is achieved by tuning fork and alternator. Reception is by Kerr cell and photographic paper."

The magazine devoted many pages to these experiments, extracts from which are reproduced below:

"A further important stage in the development of facsimile transmission over long distances was reached on February 3rd, when a demonstration of Trans-Atlantic transmission of facsimile messages by the Marconi Short Wave Beam Facsimile system was given at the Marconi Beam Receiving Station at Somerton, in Somerset. This demonstration was given to a number of newspaper correspondents representing the principal English newspapers and technical publications, who were greatly impressed by the excellence of the pictures transmitted from the Beam Station at Rocky Point, Long Island, New York, and received at Somerton. Not only did the newspapers devote a large amount of space in their news columns to appreciative descriptions and illustrations of the demonstration, but many of them gave leading articles to the subject.

A striking example of a Marconi facsimile transmission received at Somerton from the United States

The *'Morning Post'*, for instance, said in its leading article that the Marconi Facsimile system has an advantage both in certainty and secrecy: the handwriting is evidence of authenticity, and the message could not be intercepted with the case that Morse can be picked up. For the system is as great an advance on Morse in the intricacy of its working as the higher mathematics on the tally of a shepherd. Instead of the simple alternation of short-and-long, by which man has been accustomed to telegraph his messages, the varying resistances of a photo-electric cell to light are used to modulate the wireless transmitter and are reproduced in synchronism at the other end.

Not a signal merely, but an image is telegraphed. Those varying resistances to light, which, in combination, make what we see of things, are flashed with such rapidity that the picture rolls off the cylinder thousands of miles away as it is exposed to the transmitter. Here, then, is one more development of a new economy of science which is revolutionising, if not annihilating, time and space.

Commendation from Technical Publications

'The Electrician' and the *'Electrical Review'*, the two principal English electrical periodicals, also published appreciative editorial comments.

'The Electrician' says: "Our first acquaintance with this system was some two years ago, when the Marconi Company was thinking of distances in yards and obtaining results considerably less definite than those now possible over 3,000 odd miles. The progress made in so short a time is nothing short of remarkable, and is indicative of the

enthusiasm which has been brought to bear on the problems involved in the field of electrical research."

The *'Electrical Review'* writes "That photo-telegraphy has long outgrown the experimental stage is clearly indicated by the extensive and varied use that is made of the regular commercial services that are available to the public in this country, on the European Continent, and in America; indeed, last Sunday's demonstration of Marconi's Wireless Telegraph Co., Ltd., suggested that in time facsimile services may largely replace Morse telegraphy on busy circuits, and that the receipt of photo-radiograms, i.e., exact replicas of the senders' hand-written originals, will be the normal procedure."

Elsewhere in this issue a brief outline of the methods employed will vindicate the claim that they differ from others in use at present: the apparatus is certainly ingeniously contrived and reproduces beautifully clean facsimiles of the originals.

Moreover, not only has it been specially adapted to operate on short-wave beam radio circuits, thus making it possible to reduce considerably the period of transmission, as compared with long-wave services, but it is also learned with interest that the possibility contemplated of devising a system of superimposing on one and the same beam radio circuit several telegraph channels, one for facsimile transmission, and one for telephony; in fact, a triplex circuit. Thus is the value of research clearly demonstrated, and the Company is to be congratulated on having obtained results of so high an order of merit."

These independent commendations both from the lay and technical press made after a personal observation of the apparatus at work and of the results achieved are a valuable corroboration of our own claim that we have achieved something far ahead of anything previously developed in facsimile reproduction by Wireless.

How the Reporters saw it

One or two other extracts from newspaper reports may be interesting as giving a glimpse of the demonstration as seen from the outside observer's view point. The *'Daily News'*, for instance, wrote:

From our Special Correspondent. Somerton (Somerset), Sunday:

"In a 'dark room' at the Marconi Beam Receiving Station here this afternoon I watched a message come through from the other side of the Atlantic in the handwriting of the *'Daily News'* New York Correspondent.

All I saw in the dim, red glow was a spot of light running round a cylinder, but three minutes later, when a piece of photographic bromide paper was taken from the cylinder and put into a developing solution the message appeared almost immediately.

The total time that elapsed between the transmission of the message from Long Island and its appearance in facsimile here was only 3½ minutes.

The demonstration marks the opening of what promises to be a new era in wireless transmission. Facsimile messages and pictures have been transmitted across the Atlantic by the long-wave system since 1926, but the development of the Beam system demonstrated here today is quite new, and opens up many possibilities. It is far quicker than the existing long-wave system. The message that took only 3½ minutes to come from New York this afternoon would, I am told, have taken 2½ hours by the long-wave system.

This increase in speed, it is claimed, means that facsimile transmission, which, which up to the present has been of the nature of a luxury, will become the normal thing.

Eventually, I am assured by Marconi experts, there is no reason why the ordinary suburban telegram sent, say, by a husband delayed at the office, should not be delivered in the handwriting of the sender."

The *'Daily Sketch'*:

In the ruby light of the dark room of the Marconi Beam Wireless Station here this afternoon I watched a beam of light travelling round and round a cylinder of bromide photographic printing paper. Three minutes later I saw the paper removed from the receiving machine and immersed in a bath of developer, and within a few seconds I was able to read the message sent to the *'Daily Sketch'* by Gertrude Lawrence from New York.

I am leaving today for Hollywood to make audible moving pictures and I am glad to accept the invitation of the 'Daily Sketch' to radio greetings to my dear friends at home in this unique manner. Hope to see you all again soon.

Sincerely

Gertrude Lawrence

The facsimile message from Miss Gertrude Lawrence, the well-known British revue star, sent from New York to England for the 'Daily Sketch' by the Marconi wireless facsimile system.

The message was in the famous actress's own handwriting and the reproduction was perfect. A whole page of a New York newspaper was sent by facsimile transmitter during recent experiments in thirty minutes, and every word on the printed page was readable after the sheet of bromide paper on which it was received had been developed."

'North Mail' - By 'Candidus':

"At last we have real telegraphy. Ordinary telegraphy does not quite justify its name, for telegraphy means writing from a distance and the telegraph that all of us know sends the message indeed, but not the writing. But the message from Miss Lawrence in New York, which was printed yesterday, was an exact facsimile of her own handwriting.

"I leave the scientific aspects of this new discovery to those who understand them; I do not. But I am interested in the practical consequences of this new method. It means that if I did not care how much it cost I could get a letter - not a message but a facsimile letter - across the Atlantic in less time than it takes the Post Office to carry it from, say, Hampstead or Wimbledon to the City of London.

Does it not follow that the political effects of distance have now been annihilated, such a system which annihilates distance, as this does, has a political value to the British Empire that can hardly be exaggerated. For the receiving purpose of sending and receiving messages, Melbourne is now not further off than Edinburgh.

If there is an elaborate difficulty which needs careful explanation, it can be done promptly by the new wireless system, under which half a page of newspaper can be sent as easily as an ordinary inland telegram."

The *'Daily Express'*:

"Strube's Little Man did a lightning sprint across the Atlantic this afternoon. He travelled from Rocky Island, New York, to the new beam Marconi Station here in Somerset - a distance of 2,800 miles - at a speed of 56,000 miles an hour. The Little Man's time for his great Trans-Atlantic flight was only 3 minutes.

Strube's Little Man, who crossed the Atlantic for the 'Daily Express' in three minutes during the Marconi facsimile demonstration.

Strube drew the cartoon of the Little Man especially for this historic feat. The cartoon was despatched by liner to America. It was taken to the Marconi transmitting station at Rocky Island and flashed across the Atlantic. The transmission was perfect. The wirelessed cartoon was absolutely identical in every respect with the original, a copy of which I held before me for comparison."

The *'Daily Herald'*:

The drawing here reproduced by our cartoonist, Lance Mattinson, was sent to New York for transmission 6 days in advance. It took just 10 minutes to come back."

Lance Mattinson's cartoon received at Somerton for the Daily Herald during the Marconi facsimile demonstration.

Whilst these facsimile transmissions were an undoubted success, it was the understanding of radio propagation which a group of Marconi engineers under the supervision of Mr. Thomas L. Eckersley, which proved to be vital in future development of short-wave radio.

Thomas Lydwell Eckersley (1886-1959)

Detailed and comprehensive papers were written on the subject and published in numerous technical journals and magazines, but an overview of the observations which appeared in the March/April 1932 of *'Marconi Review'* gives a good precis of what was discovered:

"The attempts made by the Marconi Co. to introduce high speed facsimile between Somerton and New York in July, 1928, brought to light the existence of a type of very quick echo having a lag on the main signal which may be anything from 0.0007 to 0.0008 seconds.

These so-called echoes have a strength comparable with that of the main signal, and cannot be separated from it. In fact it seems clear that they really form part of it, being due to rays which reach the reflecting and refracting layer at a slightly different angle to those which constitute the main signal.

There are in addition echoes which are very rarely observed, having a period from one to two seconds up to several minutes, the cause of which is still unknown. They are of scientific interest only at the moment and need not be discussed here, as they affect commercial working no more than does an occasional atmospheric. Of these various types of echo, the echoes due to scattering and the echoes due to multiple reflections are the most serious in their distortion effects on the shape of high speed signals, whether the transmission considered is telegraphy, facsimile or television.

If we are to understand the conditions that produce these types of echo, we must study the manner in which the electro-magnetic waves are propagated through the medium and the general characteristics of the medium.

We know for instance that if an aerial is radiating about 1 kW, on a wavelength of about 20 metres during daylight, the field strength falls off quickly with distance, so that at about 100 km away, telegraph signals may be too weak to read, but if we continue to increase the distance from the transmitting aerial we find that the signals come on again at about 800 km, and increase in strength for another 100 km or so before commencing to gradually fall off once more.

By using a frame aerial receiver, it can be shown that the signals up to 40 or 50 km are carried by rays which travel almost parallel to the earth, and therefore they remain in the lower atmosphere, whereas those which are received at 800 km and over are conveyed by rays which reach the earth from the upper atmosphere at an angle of 20 to 30 degrees with the ground. In this case, propagation is said to take place by 'direct ray' for the first 40 or 50 km, and by 'indirect ray' beyond 800 km, and the range between 40 or 50 km and 800 km, where signals are either weak or unreliable, or vanish altogether is called the 'skip distance'."

Such understanding of radio propagation was of great importance to the design of aerials, and the judicious use of frequencies and times to reach various parts of the globe."

An article from the November 1929 edition of the same publication provided a detailed overview of the Somerton Receiving Station at the time:

"Somerton is complementary to Dorchester, being a multi-beam receiving station, with two lines of masts supporting 'Franklin' aerial and reflector systems, connected by shielded feeders with the receivers. There are now 10 short-wave receivers installed, and it is intended to increase this number to 16.

The undulators of the check circuits are mounted on the bench along the middle of the room. To economise operators' time, the tape from the undulator checking one service is fed through the undulator of a second service, so that the two records appear one above the other on the same tape. Four services can thus be recorded on two tapes, and unless one of the services is really working badly and requires close attention, it is possible for one operator to supervise both tapes and therefore four services.

In this station, all the feeders are effectively screened, and the change-over of receivers to the different feeders coupled to the various beam aerials is carried out in the specially screened feeder distribution box. A new box is being fitted to accommodate 24 aerials and 16 receivers. This feature of changing aerials is a very necessary one, and may take place a dozen times a day.

New York for instance may be working on six or seven transmitters, some of them running on idle tape until they are required for service, and in order to keep a watch on the signal strength of all of them in case it becomes necessary, due perhaps to faulty reception on one circuit, to ask them to go over to another, or to advise them whether signals are strong enough for traffic when it is offered on a transmitter not yet so employed, several aerial-receiver changes may be necessary.

The Somerton receiving area, showing eight shortwave receivers.

Owing to the thorough screening and the efficient bonding to earth of all those parts in the complete receiver equipment which carry out the function of shielding the circuits, no trouble is experienced from key clicks, and relays are in use wherever necessary."

Staff at the station often had interests in the development of radio communication, but one engineer seemed to take his interest too far, as reported locally on 29th March 1930:

Frederick Miles, an engineer, of Berwyn Bungalow, South Somerton, was fined £1 and costs for unlawfully working apparatus for wireless telegraphy without a licence between 1st October, 1929, and 26th February, 1930.

William David Campbell, Post Office overseer, of Yeovil, prosecuting on behalf of the Postmaster-General, said he had been instructed to state that these were causing a lot of trouble. Several similar cases had occurred in the immediate neighbourhood of Somerton, and proceedings taken did not seem to have had the desired effect. People still used wireless apparatus without licences.

The Postmaster-General had been put to considerable expense in finding out offenders and obtaining information against them, and he asked that the Bench would take that into consideration, when dealing with the case before them.

Vincent John Keefe, records official, stated that he called at Berwyn Bungalow on February 25th, to interview Miles, but found he was not at home. He proceeded to the Wireless Station at South Hill, where he saw the defendant and asked him if he had a wireless set installed. He replied

"Yes," and in answer to a question as to whether he had a licence, he said "No." Defendant then made a statement (which was read in Court), in which he stated he had not taken out a licence because he made sets for sale, and did not think one was necessary. Witness added that the defendant offered him 10s. for a licence, which he accepted without prejudice to the present proceedings.

Mr. Watson: "And this man works at the Wireless Station, doesn't he?"

Mr. Campbell: "Yes, Sir."

In reply to a question from the Bench, the witness Keefe said information that the Postmaster-General had received led them to think the defendant had a wireless set installed. Also, he had an outdoor aerial erected."

Somerton was one of the stations visited by delegates of the Empire Press Conference in early 1930, and contact was made with Guglielmo Marconi himself during the visit. The *'Marconi Review'* of July 1930 reported the event:

"One of the most interesting functions associated with the recent Empire Press Conference in London, which was attended by journalists from all parts of the British Empire, was the visit paid by a number of the delegates to the Marconi short wave Beam transmitting and receiving stations of Imperial and International Communications Limited at Dorchester and Somerton. While the delegates were at the Somerton station they had the novel experience of talking with Marchese Marconi who spoke from his yacht *'Elettra'* off Genoa, 1,000 miles away. The telephone conversation had been arranged so that Marchese Marconi could offer a welcome to the visitors. Messages were first exchanged between the yacht and Admiral of the Fleet Lord Wester Wemyss, and then the company, some of whom listened through headphones while others assembled around a loudspeaker, heard Marchese Marconi explain that while he was a long way from Somerton he was glad that radiotelephony afforded him a rather novel opportunity of greeting members of the Empire Press Conference together with some of his own fellow directors. Sir Augusto Bartolo, of Malta, expressed the thanks of the delegates and had a conversation in Italian with Italian journalists on board the 'Elettra'."

Delegates to the Empire Press Conference at the Somerton Beam station. Mr. T. Dunbabin (Australia) is speaking to Marchese Marconi on his yacht "Elettra" off Genoa. Admiral of the Fleet Lord Wester Wemyss is on the speaker's left.

Further tests and refinements at the station took place over the following few years, details of which were recorded in the Journal of the Television Society in December 1931:

"When the 1928-1929 observations on picture transmission between Somerton (England) and New York were used as a basis to estimate the best wavelength in 1930-1931, the waves selected proved to be too short, especially during the night. A set of accurate measurements over 12 months, up to September, 1931, confirmed the conclusion that at the present time of fewer sunspots there is a progressive increase each year in the values of short wavelengths which give best results. From March to June, 1931, a new group of echo tests in picture transmission on 15 metres and 30 metres took place between Somerton and Montreal, and Somerton and Cape Town which also showed that the density of the charges has fallen considerably from the 1928-1929 values.

In the evening the echo is strong, whereas in the early morning hours when the effects produced by the sun are at their lowest the signals are almost free from echo, there is distinctly less reflection. The shorter the wave the weaker the echo. There is also less trouble on the Somerton/Cape Town path which is mostly over land (9,500 km) than on the Montreal journey (5,500 km)."

In February 1934, the station became the receiving centre for the beam radio link to China, as the *'Marconi Review'* reported:

"A new Marconi short-wave Beam telegraph service between China and Great Britain was opened on February

3rd, providing for the first time direct regular wireless communication between the two countries. The Beam installation in China, situated at Chenju, near Shanghai, was formally inaugurated on the day of the opening of service and according to the Shanghai correspondent of the London Times 'greatly impressed distinguished Chinese and British present as an example of the combined complexity and simplicity of modern scientific work'. The corresponding stations in England are the Marconi Beam stations at Dorchester and Somerton, operated by Imperial and International Communications Limited."

Later that year, the Somerton station was involved in a historic 'first' when film footage from Australia was successfully received at the site. A report in the *"Wireless and Television Review"* from December of that year gave details:

"Starting on the 20th of October, Messrs. C. W. A. Scott and Campbell Black flew from England to Melbourne in 71 hours. Shortly after their arrival, a film showing them exactly as they arrived in the control tower of the aerodrome at Melbourne was transmitted to England. It covered the distance in less time than they took to reach Australia - 65 hours in all. At first sight of these figures one might imagine that there was nothing particularly marvellous about the achievement, particularly when one learnt that the cost of sending the film from Australia to England did not fall far short of the cost of despatching Messrs. Scott and Campbell Black to Australia. It was, however, an historic event, and time alone will enable us to judge its true importance. The ideal at which we are aiming, naturally, is that of long-distance television, which will enable such a scene to be portrayed instantaneously at any distance in this small world of ours. This remarkable milestone in radio (and film) history necessitated the transmission of 160 complete pictures over some 10,000 miles.

This film was enlarged so that each picture was of the standard size used for the transmission by radio of important news photographs. Each of the 160 pictures was transmitted separately from the short-wave station at Melbourne to the beam station at Somerton, in Somerset. At this end, each of the photographs was, of course, received separately. After the 65 hours the Gaumont British Picture Corporation had in their possession the 160 'stills' which, placed in their proper sequence, would represent a moving picture of roughly seven seconds' duration.

When one looks at an ordinary 'still' from a moving picture and notes the wealth of detail, some idea of the enormous complexity of the problem of sending not one but 160 such pictures through the ether may be gained. Add to that the fact that twelve years ago it was not possible to communicate with Australia by radio, and ten years ago it was only done for the first time on the newly-discovered short waves, and you will be getting things in their true perspective. The film was shown in 100 cinemas on the Friday night (October 26th), together with another film showing how the whole thing had been carried out. Together, they make one wonder whether the day can be very far distant when the transmission across the world of such pictures will be accomplished by television."

The occasion was also recorded locally, with more detailed financial and content information:

"A film of the arrival of Scott and Black at Melbourne, which took only three hours less to send by wireless than it took the airmen to fly to Melbourne, was shown in the news reels of London cinemas last night. At £39 a foot, the film has cost the Gaumont-British news reel, according to the calculations of Mr. Castleton Knight, who organised the whole thing, about £8,240. In addition to hundreds of pounds spent on cabling and other preliminary arrangements, it is by far the most expensive item any news reel has ever had. The sending and reception took 67 hours. The picture came by way of the Somerton beam station. A studio camera man then filmed it all over again in proper order for the news reel.

In the film Scott, unutterably weary, is seen closing his eyes in apparent slumber as he stood in the Melbourne aerodrome a few moments after the completion of the great flight. Mr. Edwards, the London owner of the victorious Comet, smokes a cigarette, and Jean Batten, the airwoman, turns laughingly from Scott to Black, who seems to make some remark to her as he takes a drink. Then Miss Batten, still smiling, turns round again to the exhausted Scott.

And that is really all there is of the first film ever to come by wireless from the other end of the world. As an achievement in wireless cinematography, it is of tremendous portent."

It became clear that the station was located in an ideal reception location, and when the stations at Skegness and Bridgwater closed it became responsible for the reception of signals on the India, Australia, South Africa, and Canada services.

In 1940, the Moorgate (London) offices of Cable and Wireless were bombed, and the operators there were relocated to Somerton for several months, dealing with traffic handled at the station and sending the message forms in bulk to London on a daily basis, by train.

In the early stages of World War 2, emergency arrangements were discussed between the Post Office and the authorities in the USA to arrange reliable back-up communication facilities in case of interruption caused by enemy action.

The United States was at the time served by the cable systems of Cable and Wireless Limited, the Western Union Telegraph Company and the Commercial Cable Company. Cable and Wireless also conducted a wireless service with New York in conjunction with the Radio Corporation of America (RCA).

The Post Office was prepared to make available eight simultaneous shortwave communication channels for westbound traffic to the United States, in addition to the two channels normally operated, and adequate receiving antennas, receivers, and terminal equipment for the reception of ten simultaneous shortwave channels for eastbound traffic from the United States. This would be provided by utilising existing spare facilities of the RCA and Mackay Radio and Telegraph Company and, if required, two further transmitters belonging to the Press Wireless Incorporation.

On 22nd November 1940 it was reported that The Ministry of Works confirmed that Cable & Wireless had applied for a licence to build a Wireless Telegraphy station at Somerton and complete one at Dorchester. This application was granted.

In the event, however, after long and detailed discussions, the proposed expansions at Somerton and Dorchester were dropped, with existing stations being adapted for this purpose.

The number of circuits continued to expand over the next few years, as well as improvements in technology. The Morse and Direct Current Cable Code (DC3) services were replaced by more efficient unprotected radio-teleprinter services, these shortly being superseded by error-correcting radio-teleprinter systems. Facsimile services, at the time the only reliable way of transmitting pictures over radio, were provided on many circuits.

In March 1947, the main aerials were damaged during a severe weather conditions and icing. 37 aerial arrays had to be repaired along with 970 'uniform' vertical elements. The repairs were completed by station staff and engineers in 69 days. The magazine *'Practical Wireless'* recalled the event in their May 1947 edition, noting that the cable network coped with the loss of the wireless telegraph circuits admirably:

"The unique advantage of flexibility enjoyed by the Empire's integrated network of cable and wireless telegraph circuits was recently vividly demonstrated. One of the two pairs of wireless stations in the United Kingdom which together handle the whole wireless telegraphic traffic to, from and through Britain were virtually paralysed by severe damage caused by the weather. Their load was, however, taken up by the cable system. The two stations are Cable and Wireless' transmitting station at Dorchester and receiving station at Somerton.

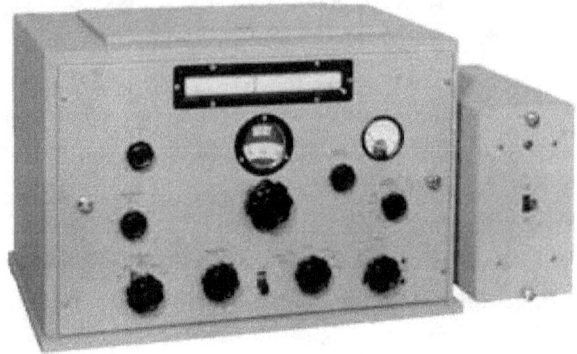

The stalwart Marconi CR150 receiver as used at Somerton and other receiving sites.

At each station, ice formed suddenly on aerials and masts. Aerials encased in cylinders of ice 3 inches in diameter crashed to the ground. Hugh icicles forming on the cross-bars and stay wires of the 100 ft. masts caused many to collapse. At one time all the beam aerials supported by the array of sixteen 300 ft. masts at Dorchester were on the ground with six 100 ft. masts and many more aerials. It has not yet been possible fully to repair the damage at either station. So effectively was the telegraph traffic taken up by the cable system, however, that the effects of the interruption of wireless circuits did not go beyond some delay to traffic with South Africa, India and the Far East."

A spectacular view of the Somerton aerial system, with crossbars atop the masts.

On 1 April 1950 the station passed into the hands of the Post Office following the Commonwealth Telegraph Act 1949, integrating the UK Radio Services of the Post Office and Cable and Wireless Ltd.

Towards the end of the 1950s, the ageing RC64, HSR and CR150 receivers were replaced by Marconi HR91, HR93 and HR24 receivers, alongside the SL60, a double-diversity ISB receiver based on the Post Office W22. The HR93 was used primarily for receiving pictures (facsimile) from overseas stations.

The SL60 diversity receivers at Somerton.

The W22 receiver was a two-channel single-sideband high-frequency radio receiver capable of rapid and unambiguous measurement of steady radio-frequencies to an accuracy equal to that of the local frequency standard, with some minor uncertainty due to the action of the counting instrument or with slightly lower accuracy for fading, noisy or drifting signals. The receiver also provided facilities for monitoring the telephony and telegraphy emissions then in general use and for field-strength measurements. The receiver covered the 4-30 MHz band only.

The HR24 was a dual diversity, dual conversion receiver receiving DSB, USB and LSB signals. Auto or manual switching of sideband was available, with Automatic Frequency Control (AFC) circuitry provided on the 2nd IF. Up to 8 telegraph channels in 3.5 kHz bandwidth or voice reception were employed. A choice of six crystal controlled spot frequencies were available, or continuous tuning over the HF band, suitable for multichannel voice frequency W/T, multichannel R/T and facsimile reception.

COMMUNICATIONS 52/255

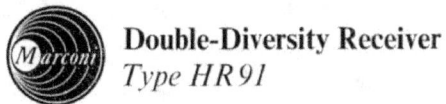

Double-Diversity Receiver
Type HR 91

THE TYPE HR 91 EQUIPMENT has a two-fold function. Firstly, it is a pre-tuned receiver providing the immediate selection of three alternative channels, and secondly, it is continuously tunable over the entire frequency range, maintaining a high standard of performance throughout. With its high degree of stability the use of the pre-tuned receiver is of considerable importance to services operating for long periods on fixed frequencies when the equipment may be essentially unattended.

The equipment is housed in two 7 ft cabinets mounted on a common base and having doors back and front.

The signal frequency panels are interchangeable as are those of the first crystal oscillators. Various arrangements of the signal frequency panels are possible such that one may serve as a spare for another, and seasonal changes can thereby be easily accommodated.

FEATURES

Immediate selection of any one of three pre-tuned spot frequencies or, alternatively, continuous tuning over the entire frequency range.

Illuminated operator's panels showing frequency in use and identification letters of station to which receiver is tuned.

Front doors may be opened to give complete accessibility of controls allowing for ease of servicing and full operating and tuning facilities. With doors closed, only controls necessary for operating on pre-set frequencies are accessible.

Panels may be withdrawn on runners giving complete accessibility to any part without involving cable disconnections.

Suitable for operation on both on/off and frequency shift keying systems. The frequency shift must lie between 400 and 1000 c/s.

DC output can be set to suit most keying systems, to operate a VF channel or telegraphy relay.

AF monitoring gives zero beat and 1000 c/s tone for on/off telegraph signals, and equal beat notes on mark and space for FSK.

Information sheet for the Marconi HR91 receiver.

355

Single-Sideband Receiver *Type HR 93*

THE INTERNATIONAL CLASS RECEIVER Type HR 93 is a dual-channel, single-sideband receiver designed to give satisfactory service under the most stringent conditions on heavily loaded long distance SSB telephone circuits. The reception of double sideband transmission is also possible.

The equipment is housed in two 7 ft cabinets mounted on a common base and having full-length doors, back and front.

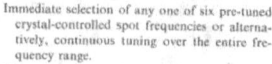

FEATURES

Immediate selection of any one of six pre-tuned crystal-controlled spot frequencies or alternatively, continuous tuning over the entire frequency range.

Illuminated panels show the frequency in use and the identification letters of the station to which the receiver is tuned.

Front doors may be opened to obtain complete accessibility to all controls for setting up and servicing purposes. With the doors closed, only the controls actually required for spot frequency working can be operated.

Panels can be withdrawn on runners giving complete accessibility to any part without involving cable disconnections.

Full metering and monitoring facilities are incorporated.

U-link method of interconnection between RF and IF units facilitates lining up and periodic circuit checking.

A local oscillator provides a signal source for lining up and testing purposes.

Provision of additional mains supply sockets for connecting external testing apparatus.

The number of valve types used has been reduced to a minimum.

An automatic frequency correction system reduces any error in the IF, corresponding to the received carrier, to zero.

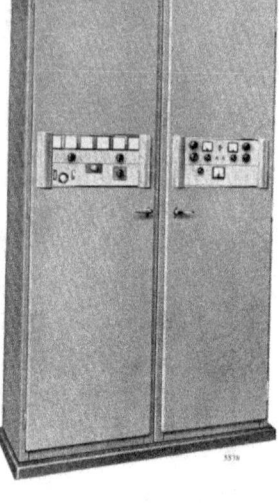

Information sheet for the Marconi HR93 receiver.

The HR93 receivers at Somerton.

Receiver banks at Somerton Radio, 1961.

The aerial and distribution boards at the station were replaced at the same time and the aerials were replaced with those of a more modern and efficient design. The old Franklin beam aerials were removed and new stacked rhombic aerials were installed, giving a greater flexibility and lower maintenance costs.

Further details were promulgated in the July 1960 edition of '*Short Wave Magazine*':

"The big radio station which many will have seen near Yeovil, Somerset, is a Post Office receiving point for overseas traffic on CW. Somerton Radio was opened in 1929, has some 50 receivers regularly in operation and, since large aerial systems are used, is on a 600-acre site.

The January 1961 edition of the '*Post Office Engineering Executive Journal*' or POEEJ, gives a detailed breakdown of the removal of the old aerials:

"A general reconstruction of the aerials and transmission lines at Somerton Radio Station, near Yeovil, has made a complete rearrangement of the aerial system necessary. Opportunity is being taken to replace the old masts, which are up to 33 years old, by masts of modern design, and this should result in a considerable reduction in maintenance costs.

The masts to be removed, 27 in number, are parallel-sided latticed-steel structures with a 12 ft. square cross-section. Twenty-one are 289 ft. high and weigh 33 tons each; the remainder are 262 ft. high and of 30 tons weight. Each mast is bolted solidly to its concrete foundation and, in addition, is supported by four stays attached at about 200 ft. from the ground. Each mast is surmounted by a cross-arm 90 ft. in length.

Somerton aerial distribution console, 1961.

There are also a considerable number of smaller masts of the multi-stayed type up to 100 ft. in height. The removal of such a large quantity of plant is a major undertaking, particularly as the full operational efficiency of the station has to be maintained throughout. Siting and erection of new masts must proceed alongside the old structures and some restrictions in the methods of recovery are therefore unavoidable. Considerable saving in costs can, however, be effected if masts can be felled and dismantled on the ground, instead of being taken down by normal dismantling methods, and it should be possible to remove 18 of the large masts by felling, without danger to other installations on site.

This antenna layout, involving 65 directional arrays, is now being rebuilt to a new plan, in connection with which 27 steel masts, many of them 280 ft. high, are being dismantled piecemeal, to be replaced by 93 lightweight stayed masts 180 ft. high. The new aerial system at Somerton Radio will require, in addition to these masts, 133 miles of wire and 510 telegraph poles for carrying the transmission lines from the antenna into the receiving building.

About 70 acres of land will be released, as the new aerial farm will cover 530 acres.

It is expected that the reconstruction program will be complete by mid-1961. Steel structures weighing in all about 1,000 tons will then have been replaced by galvanized-steel masts 187 ft. 6 in. in height with a total weight of no more than 75 tons.

The operation of felling is most spectacular, but without hazard if it is carefully carried out and spectators keep well clear. In calm weather and with the aerial system dismantled and the stays uncoupled from their ground anchorages, each mast is unbolted from the base fishplates. The structure is then free standing, but with a hawser attached to the top and to a tractor some 500 ft. away in the direction of fall. It has been calculated that a mast of the type described, unshackled and free in this manner, is safe from overturning against a uniform wind normal to a face and not exceeding 25 miles/hour.

One of the Somerton masts being felled.

The aftermath.

To allow a reasonable safety factor, felling operations are confined to periods of good weather with a measured wind speed on site of less than 10 miles/hour. A pull on the tractor hawser of about 1 ton is sufficient to unbalance the mast, and the fall is completed in less than 8 seconds. As expected, the mast falls square without distortion until it hits the ground, when considerable buckling of the structure occurs, particularly in the upper sections and cross-arm, where many bolts are broken in shear. At the end of the fall the structure moves forward bodily approximately 11 ft. along the direction of fall. At the time of writing, four of the 289 ft. masts have been felled and the recovery of the remainder is in progress."

The felling of the masts were recorded for posterity by British Pathé News, and the film (and out-takes) are readily available to view on YouTube and the British Pathé News website.

Details of staff members have been difficult to obtain, no doubt due to the high turnover of staff resulting from the various ownerships of the station. However, one well-known local employee was given a good send-off on his retirement in March 1960:

The Somerton receiving station, 1961.

"A presentation from his colleagues at the Somerton Radio Station was made on Tuesday by Mr. A. B. Pooley (engineer-in-charge) to Mr. Norman Sedding on the occasion of his retirement at the age of 65, after 40 years' service in radio communication work. He started his career with the Marconi Wireless Telegraph Co. as a ship's radio officer, and joined the staff of the Somerton Receiving Station in February 1929, two years after its opening by the Marconi Beam Wireless Service, later to become Cable & Wireless, Ltd., and eventually taken over by the G.P.O. in 1950. His farewell gift was a set of workshop accessories of his own choice. Mr. Sedding will continue to live in Somerton."

Another retirement ceremony took place in September of the same year:

"Employee Mr. Ralph Sibley, Shute Lane, Long Sutton, who has retired after 20 years as a copper smith at the Post Office Radio Station at Somerton, has been presented with a chiming clock from his colleagues. The presentation to Mr. Sibley and another of a marcasite brooch to Mrs. Sibley were made by the Deputy Engineer at the radio station (Mr. C. Anderton), who spoke highly of Mr. Sibley's work. Mr. Sibley, who was chief copper smith in charge of the aerial feeder maintenance, started his career at the station in 1940 with Cable & Wireless, Ltd., which was taken over by the Post Office in 1950."

Traffic figures at the station peaked in the mid-1960s when the main circuits employed 96 baud or 192 baud error-correcting systems using multi-channel frequency-division multiplex systems, each channel being regenerated to reduce or eliminate distortion caused by propagation disturbances.

1964 saw the use of the receiving facilities at Somerton for the Shell Company, providing data transmission both to and from selected tankers of their fleet.

The Post Office part in this experiment was to act as the carrier of data traffic to and from the ships, the signalling equipment being installed and maintained by a contractor to Shell. During the following four months the Post Office Engineering Department co-operated closely with contractors in experimental transmissions to enable Shell to choose a system. That chosen for use during the experimental period was based on the Marconi Autospec (automatic single-path error-correcting) equipment, using tone modulation of the radio transmitter at a modulation rate of 66 bauds. It was found essential, in order to minimize the effects of frequency-selective fading, to utilize space-diversity reception, and, since the conversion to direct-current signals was to be made at the Shell Centre, it was necessary to transmit signals on both diversity channels from the receiving station as tone frequencies on one pair.

This was done by modifying a path of a diversity receiver so as to shift one 'pair' of received tones in frequency, using diversity combination at Shell Centre. The service was inaugurated in July 1964 and, as an example of the amount of traffic handled, 145 messages were passed in February 1965. Calls were set up on a speech basis via the Radio Telephony Terminal (Brent) and the International Exchange, and then switched over to data transmission by each end, but the service was later controlled from the Burnham coastal radio station, which will be served by receivers at Somerton and transmitters at Portishead and will have a 4-wire private circuit to the Shell Centre.

A general view of the Somerton station, 1963.

However, towards the end of the decade, due to the movement of traffic to trans-oceanic cables and the introduction of satellite services, the number of circuits started to reduce accordingly. The Brentwood receiving station closed in 1968 and in 1971 the station at Baldock followed. This had the effect of increasing the Somerton traffic figures temporarily, and radiotelephony circuits were introduced. However, this was short-lived and the station reverted to telegraphy working.

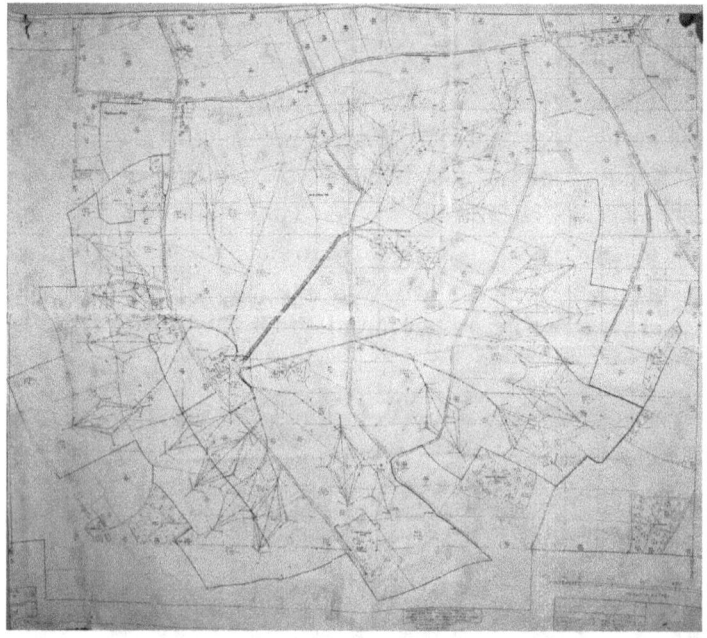

Map of the Somerton Receiving Station
showing the aerial locations.

The staff at Somerton had always been at the forefront of experimenting with new radio equipment and services, and in 1972 they came up with a way of automatically changing receiver frequency to ensure reliable reception during varying propagation conditions. The system was described as below:

"The basic method adopted is to use the distant-end transmitter to control the receiver frequency changes within a basic schedule provided by a 60-position drum-type timer making one revolution in 24 hours. This timing schedule is set by cams inserted in the drum and these operate switches which are arranged to give a 'time slot' of 48 minutes during which a frequency change to a receiver can be made.

To allow for transmitter and/or receiver frequency variations, the receiver performs a frequency searching operation (± 500 Hz at a speed of 40 Hz per second) after making a frequency change. The receiver reverts to its normal automatic frequency control locked condition when a signal of correct keying speed and acceptable telegraph distortion is recognized."

A plan to relocate the Portishead Radio receiving site to Somerton was discussed in great depth in the early 1970s. An overview of the HF Long-range radiotelegraph service written in 1970 following the closure of the 'Area Scheme' explains:

"This area scheme is now closing down because most of the overseas area stations are being withdrawn from the scheme. The remaining stations at present are Mauritius, Singapore and Cape Town. These are likely to withdraw in 1971. This means that all traffic between the UK and British ships will have to be handled by Portishead/Burnham. The estimated increase is 40% of the whole traffic (British and foreign) now handled at Burnham/Portishead.

Somerton lines and aerials, 1963.

Plans have been made, therefore, to augment equipment and staff accordingly, but unfortunately, no additional revenue will accrue to the Post Office as we already receive it under the area scheme. As an interim measure, 12 additional operating positions are being provided at Burnham and 14 additional transmitters will be used at Dorchester Radio Station to handle the additional traffic. These transmitters, although not ideally suited to the maritime service, are redundant from the fixed service.

The interim scheme will tide us over the next three or four years. The 'final' scheme envisages the building of a new receiving station at Somerton Radio Station and the closure of Burnham. Additional and more flexible transmitters will be required at Dorchester, and ultimately it may be possible to give up Portishead"

In 1972 the Union of Communication Workers (UCW) representatives from the station met with county council planning officials and Langport Urban District Council to discuss housing and associated problems in the Somerton area. It became clear that Somerton would continue to be much sought after by commuters to the Yeovil and Bridgwater areas and that property would be keenly competitive and high in price compared with similar housing in the Highbridge and Burnham-on-Sea areas. Somerton also did not offer the same social facilities offered by the current location.

The Union (under the auspices of Gerry Knott and Jim Byrne) therefore recommended that they would not make any recommendation for such a move until favourable terms of transfer had been negotiated. It was also clear that many staff would not be prepared to move to the Somerton area for family reasons and the plan was quietly dropped.

However, many magazines and newspapers got wind of the proposed move, and the December 1973 issue of *'Short Wave Magazine'* published details which proved to be somewhat premature:

"One of the world's best known long-distance communication stations, standing on a commanding site overlooking the Bristol Channel, Portishead Radio (signing some 50 c/s on numerous frequencies between 130 kHz and 22.5 MHz) is to be moved further inland to Somerton, Somerset, also a Post Office station covering a wide acreage for the erection of aerials. The function of Portishead Radio is to work ships world-wide, running from 500 watts to 15 kW on its various frequencies, and it handles an enormous volume of traffic. The move to Somerton is expected to take about four years, because in the meantime all services will have to be maintained. As a footnote, it might be mentioned that GKL always threw healthy harmonics into what, years ago before Hitler's War, was our five-metre band, enabling the South Wales amateurs of those days to find a reliable calibration point - the writer of this note can hear "VVV de GKL" even now!"

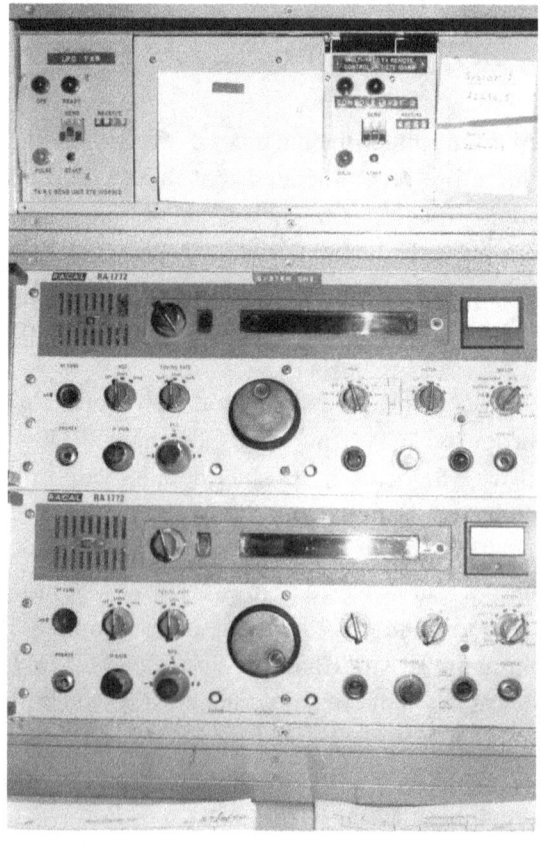

Close-up of the RA1772 receivers at Somerton.

Use of the Somerton radiotelephony services was transferred to the maritime radio service in 1976, when it became clear that the circuits at the Portishead Radio receiving station at Highbridge were insufficient to handle the demand. Six consoles were provided for this purpose and staff were transferred to the site from Highbridge by minibus on a daily basis. Prior to this, the only involvement with the maritime service was the use of the data service with Shell as described earlier, and the Autospec/Plessey circuit from the *"Queen Elizabeth 2"* liner.

New Racal RA1772 receivers were used for this service, which the company was quick to recognise in a press release from 1974.

The Racal RA1772 receiver.

"Racal's latest HF receivers have been selected by the Post Office for installation at its Somerton Radio Station, Somerset, as part of its re-equipment programme. This is the fourth in a series of orders and supplies the Post Office with ten of the Bracknell Company's receivers for use by the International Maritime Telecommunications Region (IMTR) in ship-to-shore communications. These latest receivers are immensely successful and are specially designed to meet the most stringent international specifications."

With the decline in the use of the point-to-point service at the station, some of the Somerton engineers retrained to obtain their Maritime Radio General Certificate, which allowed them to operate as Radio Officers at the Portishead station. Most of them lived locally to the Somerton site, which meant they were keen to man the radiotelephone service without the need to travel to Highbridge.

Many Portishead Radio staff recall the journey to Somerton with great amusement and affection. Many recollections were published in the author's book "Portishead Radio – A Friendly Voice on many a Dark Night" and some are reproduced below.

The daily trip could be an 'experience' as recalled by Ramsay Stuart:

"Regarding the Somerton bus - There were two drivers, Ron Westlake (aka 'Rapid Ron') being one, and the other Harry Brown. It was Harry who liked to vary the route a bit. The standard route was left at the mini roundabout on Church Street, across the railway, turn right at Watchfield, up through Woolavington, left on the Bath Road (A39) to the Pipers, right and up Walton Hill and Ivythorn Hill, right on the B3151 and right on the B3153 to Somerton.

On a June morning the view from Ivythorn Hill was spectacular. There was usually a low hanging mist and Glastonbury seemed to be an island floating in the clouds.

On one occasion Harry turned right on the A39 at The Albion, down the hill, right on the A361 to Pedwell then left and across Kings Sedgemoor to Low Ham thence through

Pedwell (where there were guinea fowl running about in the road - talk about rural - almost as rural as Nempnett Thrubwell) up to the B3153 then left and up to Somerton.

The rhynes alongside the road across the moor, lined with ancient pollarded willows, are something to behold - a double-decker bus would disappear if it fell in. It did not take but a few minutes longer that way, but it was a welcome change of scenery.

Duty at Somerton attracted a daily subsistence allowance, which was enough to buy a decent lunch and two pints of bitter at the White Hart, or the Globe next door. Better fare was to be had at the Half Moon, which was Egon Ronay rated for the cold buffet, but although the grub was marvellous it did not also run to a couple of pints.

Maritime R/T consoles at Somerton, late 1970s.

When on duty at Somerton we had the use of a vehicle to get down into town and get a meal. It was an estate car or station wagon or whatever you like to call it, but one day there was a new one, and when we drove it we were not familiar with the controls, therefore it was not unusual to see the vehicle going down the road with hazard lights flashing, head lights on, and windscreen wipers operating.

The landlord of the White Hart was a miserable little squirrel, and, although most of us liked the place, a number of incidents turned us off. I may have got this wrong, but Phil Lewis, for I think it was he, or perhaps it was Nigel Le Gresley, asked for a cup of tea, and was rewarded with a heap of verbal abuse. Most of us then transferred our custom to the Globe next door to the White Hart. The Globe was something else. It was manned at lunch-time by three ladies who were, I believe, the wives of RN personnel at Yeovilton who were the licencees.

The Globe was brilliant - you could get a 'beef burger grill' (which was a decent mixed grill but with a beef burger instead of a steak) and a couple of pints on your allowance. If you didn't eat it all, the cook would come out of the kitchen and demand to know what was wrong with it! There was an old gent who went in every day at lunchtime and had a really huge bowl of soup and a loaf of bread - I think the ladies must have quietly subsidised him. The two ladies behind the bar were overheard discussing the cook, a rather large lady. "Do you know?" one said to the other, "She came in a wrap-round dress which wouldn't have wrapped round a bloody pencil."

If a stranger came into the Globe and asked for a sandwich he would then be asked "Have you seen one of our sandwiches?" If no, then the ladies would demonstrate - you would have to have had a huge bite to get your face round it.

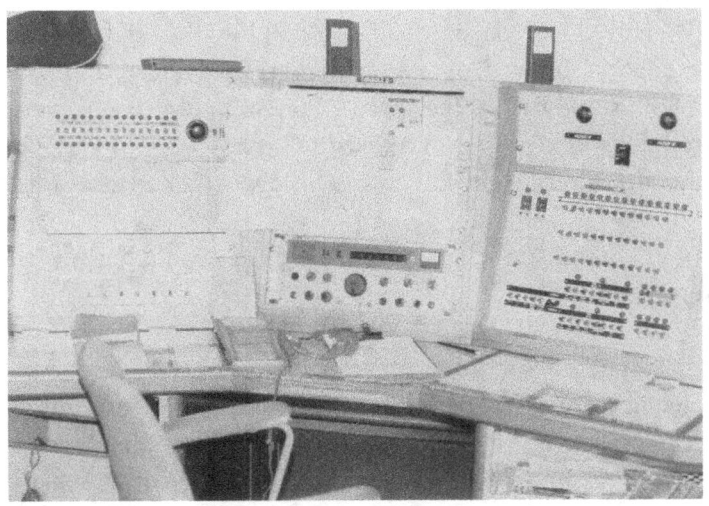

A Somerton Maritime R/T Console, early 1980s, showing the Racal RA1772 receiver.

One day, I think it was a Saturday, while the roof of the Globe was stripped for re-slating, the roof collapsed and fell into the bar. It was out of hours at the time and the ladies and their kids were in the garden at the back, so no-one was hurt. It seems that some bodger a couple of hundred years ago had made a crappy scarf joint in one of the beams. Anyway the beams were replaced with massive elm beams, thanks to Dutch elm disease which had released a huge amount of timber. There was nothing wrong with the timber - it was only the trees that died."

Sometimes the minibus would arrive back at the Highbridge site before 2300, causing one particular officious overseer to request the staff to work a few ships on W/T before the night shift arrived. These requests were 'politely' declined.

However, one particular R/O. Phil Murray, was the exception to the rule, and he regularly volunteered to take a few ships before the end of the shift. Phil was a very religious man and even refused to go home should he win the 'draw' on a night shift due to his opposition to gambling.

Frank Ryan recalls a couple of Somerton-related episodes:

"When Somerton first started operating the secondary channels on R/T, only 3 went across for the day duty. I remember with Stuart Lund and another staff member, we had an 'extended' liquid lunch in one of the pubs - we got back late to find Station Manager Don Mulholland and some Japanese gents awaiting our return. Don was not best pleased, but did not issue a 'skin', or make any further mention of this.

When an Overseer was instructed to attend for the day at Somerton, Russ Taylor used to turn up with his wife at about 9 am, hang about until 9.30, and then casually mention he was going into town - not to be seen again! Station Manager Arthur Hamblin kept on ringing up one day every half hour to try and get Russ. In the end we stopped making excuses and said he had disappeared. Never did find out what happened."

Each Radio Officer had the capability of handling ships R/T calls by selecting one of 36 rhombic aerials or one omnidirectional system, covering between 4 and 25 MHz. One of the Portishead Radio radioteleprinter circuits was a fully flexible remotely-controlled receiver, which was to be the precursor of the fully remotely-controlled service between the Highbridge receiving station and Somerton. Over 100 remotely-controlled receivers were planned to be introduced during the early 1980s.

Portishead Radio Station Manager Ernie Croskell in front of the new operational building at Highbridge, showing the microwave link to Somerton. Photograph dated 1982.

The opening of the new operations building at Highbridge signalled the end of the daily minibus trip to and from Somerton, much to the relief of many of the staff.

On 3rd April 1982 the telex line between Port Stanley and London was cut by Argentine forces, as reported internationally:

"The last telex contact between Britain and the Falkland Islands took place early today. After that the line was apparently cut by the invading Argentine forces.

Here is a transcript of the last exchange, between a subscriber in London and one in Port Stanley:

LONDON: What are all these rumours?

PORT STANLEY: We have lots of new friends.

LONDON: What about invasion rumours?

PORT STANLEY: Those are the friends I was meaning.

LONDON: They've landed?

PORT STANLEY: Absolutely.

LONDON: Are you open for traffic?

PORT STANLEY: No orders on that yet. One must obey orders.

LONDON: Whose orders?

PORT STANLEY: The new governor.

LONDON: Argentina?

PORT STANLEY: Yes.

LONDON: Are the Argentinians in control?

PORT STANLEY: Yes. You can't argue with thousands of troops plus enormous navy support when you are only 1,800 strong. Stand by please.

Then the line went dead."

Eventually, the Somerton station re-established contact with the Cable and Wireless station at Port Stanley via HF radio.

Various press reports covered the role of the Somerton station gleefully:

"Top secret messages between Argentina and occupying forces on the Falkland Islands were intercepted by British telephone engineers for a month after the invasion.

ITN reporter Michael MacMillan yesterday described how staff at the British Telecom radio centre at Somerton, near Yeovil, re-established telephone contact with the island. The Telex operator at Stanley had explained that the Argentine troops were baffled by the telephone equipment and this made it possible for engineers in Britain to eavesdrop, Mr MacMillan said.

Staff at Somerton spent two days trying to resume telephone contact with the Islands. Normally all calls from the Falklands are received at the radio centre and are then channelled to the central exchange in London, Mr MacMillan explained.

The telex operator at Stanley. Ian Stewart, yesterday relayed his first message to Britain for 10 weeks. "He said his radio equipment had been left in a filthy state by the Argentines," Mr Macmillan explained. "Most of his aerials had been broken and they couldn't be repaired because the area around them had been mined."

A spokesman for British Telecom last night said he could not comment on the claim that top-secret messages between the Argentines were intercepted in Britain.

"Communications with or from the Falklands during the time of the occupation and relating to the occupying power are a matter for the Government, not us," he said.

By 1983 the Portishead Radio receiving aerials at Highbridge had been dismantled, and the new Portishead Radio operational building was now fully operational. Racal receiver front panels were installed on each console at Highbridge, which remotely-controlled the corresponding receivers at Somerton.

Aerial selection was also remotely-controlled, with links between the two sites being provided by microwave via the Pen Hill mast on the Mendip Hills in Somerset.

An aerial direction switch on each console at Highbridge would select the appropriate rhombic aerial at Somerton. This proved successful with interruptions to the links few and far between, and remained the *modus operandi* until the service closed in 2000.

Engineers remained on-site throughout the 1990s, although the management team had offices and workshops at the Portishead Radio operational site at Highbridge.

During 1992, another attempt was made to transfer the radio services from Burnham to Somerton, which thankfully was not implemented. A report on this proposed transfer reads;

"We recommend that the principle of relocating Burnham facilities to Somerton is abandoned, but consideration given to the utilisation of omni-directional antenna receiving facilities for all services and relocation of receivers to Burnham. The requirement for Somerton as an operational site would then cease and the requirement for upgrading the power distribution, alarm systems and need for maintenance support at Somerton would then cease. The site could be moth-balled or disposed of as appropriate".

The report also mentions the problems of cost, timescales, and disruption to services, all of which would be unacceptable. In the event, the use of Somerton receiving aerials and the radio receivers was maintained until the closure of the service in April 2000.

An atmospheric view on the feeder lines at Somerton, taken towards the end of the operational service, dated 1999.

As traffic in the maritime radio service declined rapidly during the 1990s, it became clear that the days of the Somerton site were numbered. The introduction of the Global Maritime Distress and Safety System (GMDSS) in February 1999 was the death knell of the majority of the HF maritime radio stations worldwide, being satellite-based.

Some stations had already closed in anticipation of the introduction of the new service, with Portishead Radio being amongst the few who struggled into the 21st century.

The Somerton site closed at 1200 GMT on April 30th 2000 when the Portishead Radio station ceased operations. The buildings, although used for a short time by the BT 'Airwave' service until around 2005, remained dormant and unloved for many years, and are still standing at the time of writing. The aerial masts, once a prominent landmark in the area, were taken down in the early 2000s, although a couple remained for back-up or reserve purposes for a short time.

Another view of the Somerton site from around 1999. Three masts can just be seen in the mist on the right-hand side of the photograph.

In 2020, a local letting company described the site as below:

The original and distinctive Marconi style buildings are still in situ today but have been disused for a number of years and are currently boarded up (Buildings 1 & 5) Building 4 is a disused lower height former store.

All are in a mixed condition and state of repair commensurate with having been disused and derelict for a number of years.

Building 5 appears in a poor condition and has former deep service channels/voids in the floor throughout.

Building 3 is a dilapidated timber structure.

Building 2, a former workshop with inspection pit, with part profile sheeting and asbestos sheeting cladding/roof cover.

It is understood that a small housing development is being planned for the site, with the demolition of the famous buildings being considered in the planning documentation. However, no firm decision has been made on the future of the site, which remains a site of great historical interest.

Views of one of the Somerton buildings taken in 1999 showing one of the remaining masts used for microwave links and BT 'Airwave' communications.

A 2022 photograph showing the sad state of the Somerton buildings.

CHAPTER 9

AN OVERVIEW OF DORCHESTER WIRELESS STATION

It would be remiss of the author not to include details of the Dorchester station, considering its close association with the Somerton receiving station. The history of the station has been well-documented by Paul Hawkins in his dedicated publication about the station and his book 'Point-to-Point'. It is recommended that these publications be referred to for further reading. However, for the sake of completeness, an overview of the station is detailed below.

An early view of the transmitters at Dorchester.

In 1925 work commenced on the construction of the Dorchester transmitting station.

A local newspaper report from 10th January 1925 was amongst the first to announce the construction of the station:

"It is understood that the scheme for the Marconi beam station at Martinstown Corner, Dorchester, has now been definitely settled and all agreements signed. The work will be commenced forthwith. Arrangements have been made for the supply of gravel by rail and road from the Moreton gravel pits, and the dynamos, etc., will be brought to Dorchester via Great Western Railway.

This station, which will be situated about two miles out of Dorchester, on the Bridport Road, will be a beam station with a wavelength of 30 metres. It is believed also that the permanent staff at the station will be about 30, and that the work of building the station and erecting the 300 ft. steel masts will employ a large number of men for some months. This will be the first station of its kind in the United Kingdom. The site will overlook the monument to 'Kiss Me' Hardy, Nelson's Flag Captain on Blagdon Hill."

An early report on progress on the station was published in *'Wireless Weekly'* on 15th July 1925:

"Mr. E. W. Mathias, the Marconi engineer in charge of the construction of the new Beam Wireless station at Dorchester for communication with North and South America, has stated that there are at present five masts at Dorchester in a straight line at right angles to the direction in which communication is to be established - namely, New York.

The masts are 277 ft. high, each having a cross arm at the top measuring 90 ft. from end to end. The weight of the masts is approximately 45 tons each, including the cross arm, the stays being half a ton each, and 180 cubic yards of concrete were used for each of the masts.

The station would be in direct communication by land line with Radio House, the telegraphic office of the Marconi Company in London, and would be operated from there. The receiving station would be similarly connected to Radio House, so that the outgoing and incoming signals would be transmitted from and received at the same table, giving true duplex working. An important factor was that only stations situated within the angle of the beam were enabled to receive the signals."

A reporter who visited the Dorchester station in December 1927 was excited to see the station in action, as his report clearly shows:

"In a silent room, 30 feet square, on a wind-swept Dorset moor, I watched today's messages being transmitted to New York and Rio de Janeiro at the rate of 200 words a minute. It was the transmitting station of the Marconi Beam Wireless service to North and South America, and with a power of only 15 kilowatts, messages can be flashed round the world in one-seventh of a second. Outside, seven steel towers feet high carry vertical aerials. Behind them are the reflecting wires, which are a central feature of the beam system, concentrating the wireless waves into one direction.

In the transmitting room, 10 valves, hidden in oil-baths, glowed faintly. The small 'Brr-brr-rrr' of the tiny transmitting relay sending machines, automatically controlled by land lines from London, was the only sound. A small metal tongue in the sender vibrated rapidly as an operator in London fed messages into the circuits. Ammeters flickered, and in New York, 3,000 miles away, a second operator read off the messages. The current passing into the aerial was between eight and ten amperes less than is used on some motor car lighting circuits.

"We are using a 16-metre wave for America," Mr R. N. Vyvyan, engineer-in-chief of the Marconi Co., told me, "and 24 aerial wires in each bay. Behind each aerial wire is its reflector wire, placed one quarter of the wavelength away. This makes the wire send back the wave in the direction of the beam in phase with the other part of the signal, which has not been reflected. At each end of the bay the reflectors are brought round to prevent the beam spreading.

We have had some strange experiences. The other day conditions were exceptionally good. Then we found there was blurring on the tapes. We investigated, and discovered the message was repeating itself. It had gone right round the world and we were receiving it twice. Then we even got a third reception, each a fraction of a second later than the other, so that the message went round the world two and a half times.

Messages from North and South America and Egypt are received at the new Marconi station at Somerton."

The station at Dorchester opened for traffic on 16 December 1927 carrying the New York circuit, followed by services to Japan and Egypt by the end of 1928. Services to Nairobi and Teheran followed over the next few years, and the South African service was transferred to Dorchester from Bodmin.

A circuit to China was opened in 1934, with the Chinese station being located at Chenju, near Shanghai. The Australian service was relocated to Dorchester when Tetney closed in 1940.

The February 1928 issue of *'Modern Wireless'* gives a concise overview of the station at the time:

"At Dorchester there is a row of five masts, so arranged that the great circle bearing on the North American and Egyptian stations is at right angles to the line of masts. Three bays on one side of the line of masts are used for transmission to North America and two bays on the other side will be used for working to Egypt, the Egyptian transmitter being arranged for working on two alternative wavelengths.

The two active aerials for the Egyptian service and two of the active aerials for the North American service are built with a reflector common to both and placed between them.

There are two other masts, which are erected in a line at right-angles to the great circle to South America, and these carry the aerials and reflectors for the transmission of signals to Rio de Janeiro and the Argentine. Two more masts are in the course of erection, and will be used to suspend the aerials and reflectors for transmission to Japan."

*A 1928 view of the transmitter room showing the
Marconi SWB1 transmitters.*

The early transmitters were Marconi Short-Wave
Beam SWB1 models with an output power of 11kW. A
second transmitter hall with 8 of these models were added
later to improve the coverage on each circuit and to open
new services to Bangkok, Nairobi and Teheran. Manual
changeover links in the aerial feeder system were installed
to enable a change of aerial to be made when required.

Marconi himself visited the station in August 1929,
which was duly reported:

"The famous inventor, the Marchese Marconi and his
wife, the Marchesa Marconi, arrived on their wonder yacht
'Elettra' at Weymouth on Friday evening, and early the
following day they motored to the Dorchester beam station.

Here they were met by the. Engineer-in-Charge (Mr. P. J. Woodward), who conducted the party of visitors over a 90-minute tour of inspection of the station and its environs, including a number of dwellings for the leading members of the staff.

The station is now in communication with Japan (one service), New York (three services), Egypt (two services), and South America (two services).

Later the Marchese and Marchesa left for inspections of the Bridgwater and Somerset beam stations."

Presumably 'Somerset' meant 'Somerton' in this report.

In 1938 new SWB10 transmitters were introduced, which provided a 25kW power output. Later, further beam aerials were erected to take the South African services from Bodmin, and in 1941 the Australian service from Grimsby.

A photograph of the Dorchester station from 1942 showing the camouflage covering.

The Dorchester aerial array, 1963.

During 1943, Dorchester 'B' Building was added with 5 SWB10's, which could be remotely controlled from the main station.

The Commonwealth Telegraphs Act of 1949 ensured that the UK assets of Cable & Wireless Ltd. and the services and staff were passed to Post Office control, first, in 1950, to the Engineer in Chief's Department, and later to the External Telecommunications Executive (ETE) at its formation on 1st October 1952.

In 1959, 4 STC-type DS13D transmitters were installed in an extension of Dorchester 'B' Station, for use on high-traffic routes such Barbados, Lagos and Karachi, where up to 18 telegraphy channels could be required simultaneously. The transmitters, rated at 30 kW peak power could be wave-changed to any of 6 pre-assigned frequencies in 30 seconds from remote control panels in the main station.

A early general view of the Dorchester site.

The high-speed Morse systems were converted to teleprinter, and the former single-channel transmissions were upgraded to multi-channel systems, with new international services added to the station's growing list of destinations.

The station's buildings were also upgraded and extended, with new workshops added along with new canteen facilities.

The early 1960's saw the replacement of the beam aerials, their high maintenance costs, susceptibility to storm damage, and fixed frequency working led to their replacement by Rhombic broadband aerials, which could be tuned to various frequencies as required.

In 1966 three Associated Press Services began at Dorchester using Log Periodic aerials erected on the south east corner of the site, for the Eastern, Middle East and South African circuits. The latter route closed in 1975.

Because of the success of satellite communication and the provision of trans-oceanic cables, all point-to-point services had ceased or were transferred from Dorchester by 1970. The remaining SWB1 transmitters were removed and the 'B' Station SWB10's transferred to the main station, the complement then being: 11 SWB10, 2 SWB8 and 4 DS13D transmitters.

The Dorchester main building and aerials, 1963.

After 1971, apart from the continuing Press services, all the transmitters were used for long range maritime telegraphy services, using Stacked Quad Dipole aerials for the 8, 12, 16 and 22 MHz maritime bands and vertical bi-cone radiators for the 4 and 6 MHz maritime bands. They formed part of the world-wide radiotelegraph service established through Portishead Radio.

The HF rationalisation scheme, which was then being implemented, eventually concentrated maritime telegraph services at Rugby and Ongar stations. Here, more modern versatile transmitters were available, following the transfer of many point-to-point services to satellite working via Goonhilly Satellite Earth Station.

The station also provided services for the press and for maritime use, being one of the transmitting sites for the Portishead Radio HF maritime service. However, by the early 1970s, the overseas services had more or less been replaced by satellite links, leaving only the press and maritime traffic as the only remaining services.

During the 70's these services were relocated to sites at Ongar and Rugby under the Post Office rationalisation scheme, and in 1977 the station celebrated its 'Golden Anniversary' by opening the station to the public to see the station in operation and to organise an exhibition of radio communications. Sadly, the days of the station were numbered, and the Dorchester station closed during 1979, with other UK transmitting stations closing around the same time.

The station buildings were left unloved for many years after the closure; the surrounding land returned to agricultural use, and the transmitting equipment removed and subsequently scrapped. However, from 1984 the station buildings were taken over by the Friary Press and expanded to cater for the needs of the business.

At the time of writing, the original transmitter building is still standing, along with the expanded buildings mentioned earlier, and the complex is now used for industrial use by Advantage Digital Print.

The Dorchester site, early 2000s.

No evidence of the aerials sadly remains, although one mobile telephone mast has been erected close to the building. On the other side of the A35 is a row of houses known as 'Radio Station Houses' which were originally built as accommodation for staff in the 1920s.

APPENDIX 1 - BRIDGWATER BEAM WIRELESS STATION – EQUIPMENT

AERIAL SYSTEM

Standard copper tube 'feeder' system for eight bays includes:

All ground work, viz:

- Transformer Boxes
- Coupling Boxes
- Expansion Joints and Joint Boxes
- Two 5-mast Single Way Beam Aerials.

RECEIVER ROOM – FOR THE SERVICE WITH SOUTH AFRICA

Receiver unit, comprising:

- 2 Feeder Terminal Boxes
- 1 L.F. Modulator
- 1 1st Het.
- 1 H.F. Amplifier
- 1 1st Detector
- 1 H.F. Amplifier
- 1 H.F. Amplifier
- 1 2nd Detector
- 1 D.C. Bridge
- 1 Control Unit: Fitted with Test Instruments (Weston)
- 1 Iron Frame for all above
- 1 Twin Testing Cord with Clips and Plugs
- 1 Twin Bridging Cords with Clips and Plugs

- 1 Pair Telephones, 60 plus 60 (Sterling)
- 2 Wood H.T. Battery Boxes
- 4 42v H.T. Batteries (Ever Ready Type W.561)
- 2 42v H.T. Batteries (Ever Ready Type W.561s)
- 1 Short Wave Het. Coil
- 1 Long Wave Het. Coil

FOR THE SERVICE WITH CANADA

Receiver Unit: Equipment similar to that detailed above for the service with South Africa.

FOR THE SERVICE WITH SOUTH AFRICA

Recording Apparatus:

2 High Speed Undulator Recorders Nos. 245725 & 277301 comprising:

- 1 Two-way Socket
- 1 Three-way Socket
- 1 Two-pin Plug
- 1 Three-Pin Plug
- 1 Tumbler Switch with two-way Socket

2 High Speed Tape Pullers, Nos. 324591 & 3124673 comprising:

- Motor 110 volt
- 1 Two-way Socket
- 1 Two-pin Plug
- 1 Regulating Resistance (in base)

1 Automatic Telephone Coy's High Speed Relays No. 552 & No. 547:

1 Base for, fitted to: 2 1mf Condensers, 4 200 ohm Resistance Coils (anti sparking device)

- 1 Galvo (G.P.O. make) No. 1a
- 2 Lamps 110v 60 watt
- 2 Porcelain Bases for lamps
- 1 Galvo (G.P.O make) No. 1a
- 1 Resistance Box (Marconi make)

Speaker Apparatus: Sounder.

- 1 Sounder (G.P.O make) 20 ohm
- 1 Sounder hood
- 1 D.C. key (5 terminal), fitted to: 2 2mf Condensers, 2 200 ohm Resistance Coils (Anti Sparking Devices)
- 1 Relay "B" fitted to: 1 2mf Condenser, 1 200 ohm Resistance Coil (Anti Sparking Device)
- 2 Galvos. No. 1a
- 2 Lamps, 110v 8 cp. Carbon Fils.
- 2 Porcelain Lamp Holders
- Leclanche Cells

FOR THE SERVICE WITH CANADA

Recording Apparatus:

2 High Speed Undulator Recorders Nos. 328765 & 288263 comprising:

- 1 Tumbler Switch
- 1 1500 ohm, 0.125 Amp Resistance
- 2 Three-way Sockets
- 1 Three-pin Plug
- 1 Two-pin Plug

2 Tape Pullers, Nos. 324579 & 324255 comprising:

- 1 Motor 110v.
- 1 Three-way Socket
- 1 Three-pin Plug

2 A.T.M. High Speed Relays Nos. 571 & 604

1 Base for, fitted to: 2 1 mf Condenser, 4 200 ohm Resistance Coils (anti sparking devices)

APPENDIX 2 – OFFICAL SOUVENIR PROGRAMME OF THE IMPERIAL CONFERENCE DELEGATES TO THE DORCHESTER AND BRIDGWATER BEAM WIRELESS STATIONS, 8TH NOVEMBER 1926.

Souvenir Programme

of

Visit of Imperial Conference Delegates

to

Dorchester and Bridgwater Beam Stations

Saturday, 8th November, 1930

CABLES AND WIRELESS LIMITED

IMPERIAL AND INTERNATIONAL COMMUNICATIONS LIMITED

CABLES AND WIRELESS LIMITED

President
THE RIGHT HON. LORD INVERFORTH, P.C.

Governor and Managing Director
J. C. DENISON-PENDER, ESQ.

Deputy Governor and Managing Director
THE RIGHT HON. F. G. KELLAWAY, P.C., J.P.

Deputy Governor
THE RIGHT HON. THE EARL OF MIDLETON, K.P., P.C.

IMPERIAL AND INTERNATIONAL COMMUNICATIONS LIMITED

Chairman
SIR BASIL P. BLACKETT, K.C.B., K.C.S.I.

Managing Directors
J. C. DENISON-PENDER, ESQ.
THE RIGHT HON. F. G. KELLAWAY, P.C., J.P.

COURT OF DIRECTORS OF ABOVE COMPANIES IN ADDITION TO THE FOREGOING

SIR CHARLES STEWART ADDIS, K.C.M.G.
F. R. S. BALFOUR, ESQ., D.L., J.P.
SIR CHARLES COUPAR BARRIE, K.B.E., D.L.
SIR F. J. BARTHORPE, H.M.L.
THE RIGHT HON. THE EARL OF BESSBOROUGH, C.M.G.
COL. THE HON. A. G. BRODRICK, T.D., A.D.C.
THE RIGHT HON. THE EARL OF CLARENDON, D.L., J.P.
ADMIRAL H. W. GRANT, C.B.
HENRY CHARLES HAMBRO, ESQ.
THE RIGHT HON. THE EARL OF INCHCAPE, G.C.S.I., G.C.M.G., K.C.I.E.
FRANCIS A. JOHNSTON, ESQ.
MAJOR HARRY LEFROY, P.C., J.P.
THE MARCHESE GUGLIELMO MARCONI, G.C.V.O., LL.D., D.SC.
JOHN F. O'MALLEY, ESQ., F.R.C.S.
THE HON. GEORGE PEEL, D.L., J.P.
LORCAN G. SHERLOCK, ESQ., LL.D.
ADMIRAL OF THE FLEET, THE RIGHT HON. LORD WESTER WEMYSS, G.C.B., C.M.G., M.V.O., D.C.L., LL.D.

General Manager and Secretary of Both Companies—EDWARD WILSHAW, ESQ., F.C.I.S.

·

403

TIME TABLE

&

Owing to the extensive nature of the day's programme, it will be necessary to keep closely to this Time-Table. Visitors are, therefore, requested to give their assistance in keeping to the times shown.

8.45 a.m.—Leave Royal Hotel, Bristol.

9.10 a.m.—Leave Temple Meads Railway Station by Special Train for Dorchester. Breakfast will be served on this train.

11.10 a.m.—Arrive Dorchester Railway Station and leave by saloon road coach for Dorchester Beam Wireless Station.

11.20 a.m.—Arrive Dorchester Beam Wireless Station.

12.20 p.m.—Leave Wireless Station by saloon road coach for Weymouth.

12.45 p.m.—Arrive Gloucester Hotel, Weymouth, for luncheon.

2.30 p.m.—Leave Gloucester Hotel for Railway Station.

2.45 p.m.—Leave Weymouth Railway Station for Bridgwater. Tea will be served on the train.

4.15 p.m.—Arrive Durston Railway Station and leave by saloon road coach for Bridgwater Beam Wireless Receiving Station.

4.30 p.m.—Arrive Bridgwater Beam Wireless Receiving Station, for wireless telephone demonstration.

5.40 p.m.—Leave Bridgwater Wireless Station for Taunton Railway Station.

6.28 p.m.—Leave Taunton Railway Station for London. Dinner will be served on the train.

9. 0 p.m.—Arrive Paddington.

NOTE—THOSE VISITORS PROCEEDING FROM LONDON BY THE 8.30 A.M. TRAIN FROM WATERLOO ON SATURDAY MORNING, WILL PARTICIPATE IN THE ABOVE ITINERARY AS FROM LUNCHEON AT WEYMOUTH.

2

TO-DAY'S JOURNEY
A Moving Picture of History

A JOURNEY through the West Country, planned to embrace some of the stations of IMPERIAL AND INTERNATIONAL COMMUNICATIONS LIMITED, brings the visitor into close contact with the early history of every member of the British Commonwealth of Nations.

When Cabot first touched the fringes of Canada, when the founder of Sydney came back to Somerset to find a quiet grave, when Clive of India recuperated his strength in Bath and Wolfe of Quebec lived there just before his last campaign, when Richard Clark set out on the great voyage to Newfoundland, when Drake and his West Countrymen first sailed round the extremity of South Africa, giving the world a new sea route—great gains were in store for the Dominions then in the shaping, and some little honour for the West Country that was associated with these men.

Aeroplanes and ancient history are among the attributes of BRISTOL. She is noted for her import trade in wool, meat, timber, and grain, as well as for what Queen Elizabeth called " the fairest, the goodliest, the most famous parish church in England."

The shrewdness of the tradesmen and city fathers of 700 years ago is responsible for Bristol and all that the port has meant to international commerce. These men bravely diverted a river to form an artificial harbour. Within a few generations Bristol was the second city in England ; she is still, with her 377,000 inhabitants, one of the most important ports in the country.

Adventure has always been her share. From here the elder Cabot sailed the Atlantic in 1497. From here the Great Western, one of the first two British steamers to cross the Atlantic, left the stocks and steamed west in 1838. Here, a hundred years ago, was planned and begun one of the wonders of the world for that time ; the Clifton suspension bridge, 1,352 feet long, 287 feet above the tidal water, and 702 feet across its central span.

3

BATH, now a fifth the size of her sister-town, was a city on the map when Bristol was but a ferry.

There is a noble bathing-pool in Bath, in which sufferers from many ailments regain health by immersion. This bath is lined with lead that was laid by Roman craftsmen 1,800 or 2,000 years ago. Beside the bath run conduit-pipes to carry water from the springs. These pipes, too, were made and laid by Romans.

And during the Great War the Roman springs, yielding 500,000 gallons a day, were doing their healing work; the people of Bath saw the distinctive uniform of Canadian, Australian, South African, New Zealand, Newfoundland, and Indian soldiers taken daily to the baths to gain relief from the injuries of battle.

A glimpse of Bath, with her terraces of stone houses rising on the green slopes of seven hills, is a constant reward for travellers from all parts of the world. Perhaps her social associations account for some of her glamour. Tablets bearing famous names appear on the walls of scores of houses. Dickens, Dr. Johnson, Sir Walter Scott are among the varied figures that have passed among the pillars of the old Pump Room.

A mile or two further on we pass, in a deep valley, between the famous hydropathic establishment of Limpley Stoke on the left and Monkton Combe, one of the better-known public schools of England, on the right. BRADFORD-ON-AVON is the delight of archaeologists. One of the finest of the very few Saxon churches in the country stands there to-day, as stout as when it was first built 1,200 years ago. The little town, with its 4,621 inhabitants, was famous for its wool centuries ago. The great Yorkshire wool metropolis of Bradford, with a population of 286,000, was actually named after this modest parent town in the West Country.

TROWBRIDGE is the administrative headquarters of Wiltshire. Cloth has been woven in the town since the sixteenth century, but nowadays Trowbridge has become an important centre of the dairy industry. Its cream and butter are exported all over the world.

WESTBURY, on the edge of the Wiltshire Downs, is noted for its White Horse, a colossal silhouette cut in the solid chalk on the hillside.

4

FROME is a quiet little town that was centuries old when the Domesday Book was compiled. CASTLE CARY, one of the most beautiful villages in England, is near the birthplace of the famous Berkeley who became Governor of Virginia.

YEOVIL is a town whose roots in history reach to the Stone Age, of whose implements it has a rich store.

DORCHESTER, the " Casterbridge " of Thomas Hardy's novels and the home of the writer for years, has the most perfect Roman amphitheatre in England, 220 feet wide. On the edge of the town stands a wireless beam transmitting station of Imperial and International Communications.

Then we come to the sea and to George III's favourite resort, WEYMOUTH. It was from here that Clark sailed on his voyage to Newfoundland with Humphrey Gilbert in 1583. The great naval station of PORTLAND is in Weymouth Bay.

Travelling back through Yeovil we turn westward to LANGPORT WEST a station from which the Somerton receiving station of the wireless beam system of Imperial and International Communications can be reached. Then comes ATHELNEY, the traditional site of the cottage in which King Alfred had his ears boxed for allowing the cakes to burn.

BRIDGWATER was the birthplace of Admiral Blake, who established British supremacy on the seas in the 17th century, in the critical early days of oversea settlement. Appropriately enough, England's most modern link with oversea land is here, in the Imperial wireless beam receiving station of Imperial and International Communications.

Finally we reach TAUNTON, in whose castle Judge Jeffreys held his Bloody Assize after Sedgemoor, the last battle fought on English soil. At this one Assize in 1685, he sent 330 prisoners to the scaffold. Colonel Kirke, leader of " Kirke's Lambs," hanged many men without trial from the signboard of the White Hart. Just outside the town is the house in which Coleridge wrote " The Ancient Mariner."

DORCHESTER BEAM WIRELESS STATION

The Dorchester Beam Station is situated two miles from Dorchester on the Bridport Road, and consists of two main buildings, one containing the plant which provides power for running the station, and the other containing seven wireless Beam Transmitters.

THE WIRELESS TRANSMITTERS.

Each wireless transmitter is enclosed in a brass framework in which the valves and other components are mounted on glass insulators, this method of mounting being necessary owing to the high-frequency electric currents utilised in short wave wireless transmission.

The principal valves used are of the oil-cooled type in which the metal anode, which is heated to a high temperature by a powerful electric current, is cooled by a constant stream of oil.

The transmitting sets are arranged on both sides of the main Transmitter Hall, each set being labelled to show the services for which it is used. In the centre of the hall, between the transmitters, are the control tables and at one end are the landline tables through which the wireless transmitters are connected to the landlines from the Telegraph Office in London.

On these tables loud speakers are so arranged that the operating staff can keep a check on the signals received from London and also on the wireless signals as they are emitted from the transmitting aerials.

THE FEEDER AND AERIAL SYSTEM.

The transmitters are connected to the aerials by copper tubes which are known as feeders. These are carried above the ground on metal supports.

There are three aerial-reflector systems in use at Dorchester. The largest, which has five masts, is used for transmission to the United States of America

MARCONI—MATHIEU BEAM TELEPHONE RECEIVER

on one side and to Egypt on the other. The second, having three masts, is used for transmission to the Far East on one side and Argentine and Brazil on the other. The third, having two masts, is also used for transmission to Argentine and Brazil. In each case the line of masts is erected at right angles to the direction in which it is desired to transmit.

All the masts are 277 ft. in height, with cross-arms 90 feet wide at the top. The cross-arms carry the triatics (connecting wires) from which the aerial wires and reflectors are suspended.

The active aerial consists of an array of vertical wires behind which is suspended another array of vertical wires which form the reflector, and accurately focusses the Beam signals on to the corresponding overseas receiving station.

WIRELESS TELEPHONE EQUIPMENT.

In the main Transmitter Hall is a telephone modulating equipment, which, by means of switch gear, can be connected to any of the transmitters so that telephony may be transmitted to the various countries with which Dorchester is in communication.

7

BRIDGWATER BEAM RECEIVING STATION
Left, five masts for receiving from Canada. Right, five masts for receiving from South Africa.

BRIDGWATER BEAM WIRELESS STATION

As can be seen from the photograph, the aerials of Bridgwater Beam Wireless Station make an imposing picture. There are two sets of masts, one arranged to receive from Canada, and the other from South Africa. Each row consists of four bays of which two are used for reception on one wave length and two on another wave length. This enables reception to be continuous both by day and night.

The station is also fitted with apparatus for telephony experiments, and the subject of beam wireless telephony has had the close attention of the Company's engineering staff. The developments in this direction have been such that it is now technically possible to telephone with great clarity by means of beam telephony to the most distant parts of the world to which beam services are available.

From Bridgwater it is hoped to arrange, during the afternoon, for members of the party and representatives of the Press to converse by wireless telephone.

The Dorchester and Somerton stations of Imperial and International Communications Limited, have also played an important part in the telephony experiments which the Marchese Marconi has carried out in testing the latest wireless telephony installation for use on board ship.

In the course of these experiments he has been in daily telephone communication with London through Dorchester and Somerton and has also spoken to Sydney, Buenos Aires, Rio-de-Janeiro, New York, Montreal, Bombay and Cape Town, thus covering practically the whole world. His conversation with Sydney was over a distance of 9,000 miles which was more than three times the previous record for telephony from a ship at sea.

The photograph of the Marconi-Mathieu Beam Telephone Receiver is shown on the previous page.

8

LUNCHEON at the GLOUCESTER HOTEL
WEYMOUTH

SIR BASIL P. BLACKETT, K.C.B., K.C.S.I.,
Chairman of Imperial and International Communications Limited, presiding.

WINES

CHATEAU LA FLORA
BLANCHE

ᔕ

BEAUNE SUPERIEUR
LEON MONTE FRERES
1919

ᔕ

COURVOISIER'S 1875
LIQUEURS

ᔕ

MINERALS.   SPIRITS

MENU

GRAPE FRUIT AU MARASQUIN
HORS D'OEUVRES VARIÉS

CRÈME CHANTILLY

SOLE A LA NORMANDE

POULET DE SURREY ROTI AU LARD
HARICOT VERTS
POMMES RISSOLÉES

POIRES DELYSIA
CHARLOTTE ST. JOSÉ

CAFÉ

SPEECH BY SIR BASIL P. BLACKETT, K.C.B., K.C.S.I.

Responses by :
THE RIGHT HON. J. H. THOMAS, M.P.
His Majesty's Secretary of State for Dominion Affairs.

THE RIGHT HON. J. H. SCULLIN, M.P.,
Prime Minister of Australia.

9

411

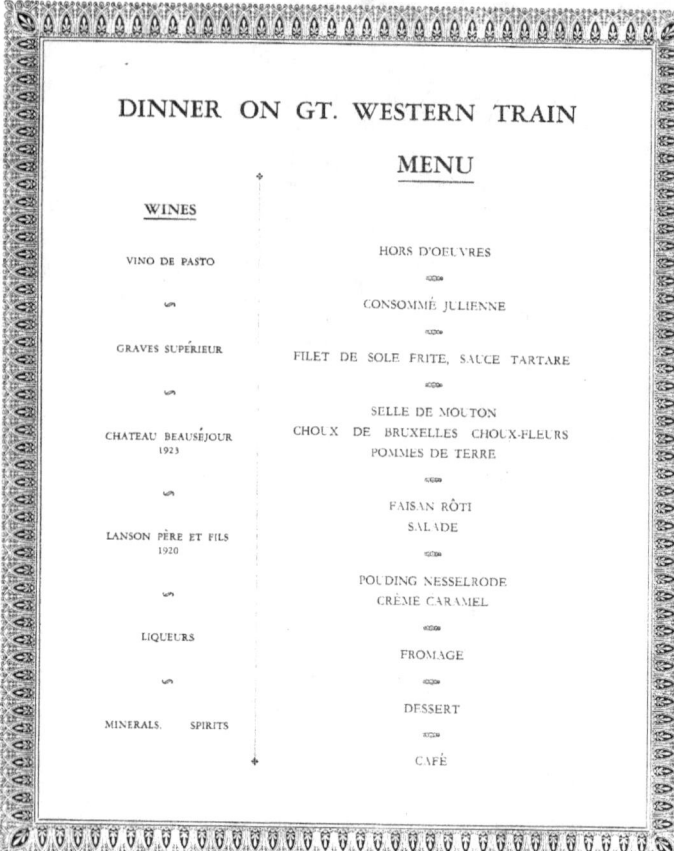

DINNER ON GT. WESTERN TRAIN

MENU

WINES

VINO DE PASTO

GRAVES SUPÉRIEUR

CHATEAU BEAUSÉJOUR
1923

LANSON PÈRE ET FILS
1920

LIQUEURS

MINERALS.   SPIRITS

HORS D'OEUVRES

CONSOMMÉ JULIENNE

FILET DE SOLE FRITE, SAUCE TARTARE

SELLE DE MOUTON
CHOUX DE BRUXELLES CHOUX-FLEURS
POMMES DE TERRE

FAISAN RÔTI
SALADE

POUDING NESSELRODE
CRÈME CARAMEL

FROMAGE

DESSERT

CAFÉ

10

412

THE ORIGIN AND PROBLEMS

of

Cables and Wireless Limited and Imperial and International Communications Limited

⊛

Cables and Wireless Limited (the holding company) and Imperial and International Communications Limited (the operating company) are the direct outcome of the Imperial Wireless and Cable Conference of 1928, at which Conference Representatives of the British and Dominion Governments and of the Government of India were represented.

The Report of that Conference prescribed the constitution, capital and assets of Imperial and International Communications Limited, determined its direction and management, specified its liabilities for assets to be transferred, and limited its profits by the imposition of a standard net revenue. It provided further for the setting up of an Imperial Communications Advisory Committee, consisting of representatives appointed by the constituent parts of the Empire, and gave this Committee specified powers of supervision and control in such matters as the increase or reduction of rates and the maintenance of strategic cables.

The formation of the Company dates from its registration in April, 1929, but actual physical and effective control of the cables was not transferred to it until September of that year, when there were brought under the Company's control 164,400 nautical miles of cable, 13 cable ships, 253 cable and wireless stations and offices, and a staff including supernumeraries, of 13,000. It took over the Cable System of the Eastern Associated Telegraph Companies, the Pacific Cable Board's cables and Wireless system, the Wireless Services of Marconi's Wireless Telegraph Company, Limited, the Imperial Atlantic Cables and, on a lease for 25 years, the Post Office Beam services.

Modern finance and business are almost completely dependent on communications. Not only do communications maintain trade and commerce in their daily needs, but they develop new sources of supply and fresh markets for manufactured goods.

The creation of Imperial and International Communications Limited was the result of a realisation that the British Empire must not only be provided with the best possible strategical communication for use in the event of war, but also with the most efficient and economical system for the rapid exchange of business, press and social messages.

This great Imperial Public Utility Company, therefore, looks confidently to the Governments of the Empire for their encouragement and support of the practice of " Telegraphing Imperially."

VISIT OF IMPERIAL CONFERENCE DELEGATES TO
DORCHESTER AND BRIDGWATER
BEAM WIRELESS STATIONS

Names of those participating in the Visit

Table No.

A.

14 Abbott, E., Esq., and Mrs. Abbott.
21 Anderson, Brigadier-General Stuart M., D.S.O.
22 Australian Press Association.

B.

20 Bale, W. H., Esq., and Mrs. Bale.
8 Balfour, F. R. S., Esq., D.L., J.P. and Mrs. Balfour.
20 Baring, J., Esq.
19 Barnett, B. L., Esq., M.C.
2 Barrie, Sir Charles C., K.B.E., D.L.
11 Barthorpe, Sir Frederick J., H.M.I.
4 Besly, J. C., Esq.
6 Blackett, Sir Basil P., K.C.B., K.C.S.I., and Lady Blackett.
20 Boyle, Alderman Percy, and Mrs. Boyle: Mayor and Mayoress of Weymouth.
1 & 8 Brennan, The Hon. F., M.P., Mrs. and Miss Brennan.
1 Brodrick, Colonel The Hon. A. G., T.D., A.D.C.
10 Brown, C. C., Esq., and Mrs. Brown.
18 Bryant, The Right Hon. Walter and Mrs. Bryant; Lord Mayor and Lady
[Mayoress of Bristol.

C.

3 Cameron, Lieutenant H. L.
9 Cameron, L., Esq.
5 Canadian Press Association.
5 Cape Argus.
5 Cape Times.
8 Collie, J., Esq.
9 Collyer, J. C., Esq.
1 Corbett, Sir Geoffrey L., K.B.E., C.I.E., I.C.S., and Lady Corbett.

2

Table No.

17 Midleton, The Right Hon. The Earl of, K.P., P.C., and Lady Midleton.
17 Moloney, Mrs. P. J.
5 Morning Post.
14 Munro, Major J. J., O.B.E., M.C.
4 Murray, Lieut.-Colonel E., O.B.E., and Mrs. Murray.

N.

5 New Zealand Associated Press.

O.

16 Observer.

P.

7 Peel, The Hon. George, D.L., J.P., and Lady Agnes Peel.
21 Penman, A. T., Esq.
15 Pope, Major A. N., and Mrs. Pope.
23 Press Association.

R.

23 Rayner-Smith, C., Esq.
21 Reuters.
23 Richards, W. G., Esq.
3 Rivett, Dr. A. C. D., and Mrs. Rivett.
12 Rudman, H. J. G., Esq., Sheriff of Bristol.
17 Ryrie, Major-General Sir Granville, K.C.M.G., C.B., and Lady Ryrie.

S.

18 Satge, Lieut.-Colonel H. V. B. de, C.M.G., D.S.O.
6 Scullin, The Right Hon. J. H., M.P., and Mrs. Scullin.
11 Steel, Lieut.-Colonel W. A., M.C.
2 Sturman, Lieut.-Colonel E. A., C.B.E.

T.

21 Taylor, C., Esq.
9 Tarver, F. H. C., Esq.
6 & 2 Thomas, The Right Hon. J. H., M.P., Mrs. and Miss Thomas.
18 Threlfall, M., Esq., and Mrs. Threlfall.
23 Times.
9 Tory, H. M., Esq., D.Sc.

3

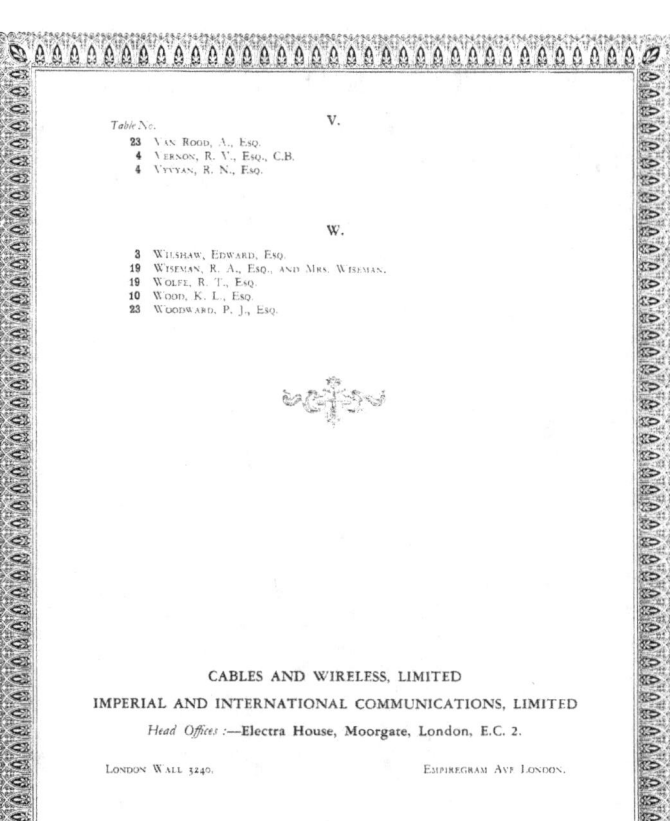

Table No.

V.

23 Van Rood, A., Esq.
4 Vernon, R. V., Esq., C.B.
4 Vyvyan, R. N., Esq.

W.

3 Wilshaw, Edward, Esq.
19 Wiseman, R. A., Esq., and Mrs. Wiseman.
19 Wolfe, R. T., Esq.
10 Wood, K. L., Esq.
23 Woodward, P. J., Esq.

CABLES AND WIRELESS, LIMITED

IMPERIAL AND INTERNATIONAL COMMUNICATIONS, LIMITED

Head Offices :—**Electra House, Moorgate, London, E.C. 2.**

London Wall 3240.   Empiregram Ave London.

4

APPENDIX 3 – LETTER FROM THE CHIEF ENGINEER OF THE POST OFFICE CONFIRMING SUCCESSFUL TESTING OF THE CANADIAN BEAM CIRCUIT.

cc.18.10.

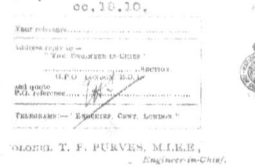

Office of the Engineer-in-Chief,

General Post Office (Alder House),

London, E.C.1.

18 October, 1926.

COLONEL T. F. PURVES, M.I.E.E,
Engineer-in-Chief.

To Messrs. The Marconi Telegraph Company Limited,
Marconi House,
Strand, London.

_____

Beam Agreement dated 28th July, 1924.

The Preliminary Certificate.

This is to certify that the sending Beam Station
erected at Bodmin for transmission to Canada and the
receiving Beam Station erected at Bridgwater for reception
from Canada have been submitted to seven consecutive days
working and after making due allowance for the period
during which an abnormal electrical storm was experienced
the stations satisfactorily fulfilled the conditions of
being capable of sending and receiving at the same time
to and from the Canadian Station one hundred words of
five letters each per minute during a daily average of
eighteen hours in accordance with clause 6 of the Beam
agreement.

Engineer-in-Chief.